RATIONAL COSMOLOGY:

OR

THE ETERNAL PRINCIPLES AND THE NECESSARY LAWS OF THE UNIVERSE.

BY

LAURENS P. HICKOK, D. D.

UNION COLLEGE.

NEW YORK:
D. APPLETON & COMPANY,
346 & 348 BROADWAY.
LONDON: 16 LITTLE BRITAIN.
1858.

PREFACE.

THERE must somewhere be a position from whence it may clearly be seen, that the universe has laws which are necessarily determined by immutable and eternal principles. Nothing in nature, and equally so not nature itself, can be made intelligible except as it has been subjected to rational principle, and such principle must both have been, and been made controlling, in the very origination of nature, or nature must forever be without meaning or end. That principle, then, to an all-perfect insight, must disclose within itself what the facts must be, and no induction of facts can at all be needed by the absolute reason.

But the finite reason, with its partial insight, must have too limited a comprehension of the eternal principle, to be able adequately to follow out all its determined results from itself, without a reference to the facts that have been determined by it to guide his intuitive processes. What already is must often help him to see what eternally must have been, and without the suggestive fact he would have failed to find the determinations of the principle. Still the mere facts in nature can never suffice to bring him to the eternal principle. No single fact, and no possible induction of facts, can give the principle; for all single facts are meaningless, and all induction of facts wholly aimless, except as some apprehension of the principle is already attained. Facts therefore are useless, and leave the insight helpless without some apprehension of the eternal principle ; but the apprehension of the principle is too inadequate to the finite reason, to permit the insight to follow out all its determinations without some reference to the facts actually determined. It will thus ever be true for the finite human reason, that with the mere facts of nature he can never rise to any science of nature, and with the partial apprehension of the principle he can never follow it out in all its necessary determinations, and hence his only sure progress must be, first an apprehension of the principle, more or less inadequately, and then a following out of the principle in its necessary laws by a reference to the actual facts that have already been determined by it. Pure principles will thus always be more clearly and completely read by the human mind where there is the most clear and complete possession of the actual facts, and the study of the principle in them and by them. The facts are nothing for philosophy except as seen to be determined in their principle, but they are much for philosophy when used by the insight for the development of the determinations of principle.

Inasmuch, then, as Nature is a rational creation, the Creator must have put his own idea into it, and the principles that determined in the making, must come out in its ongoing. The development of the determinations of the pure principle must harmonize with, because they have necessitated, the laws in the actual facts, and the study of the facts in the necessary laws, and of those laws in the determinations of their eternal principles, is the only possible method for attaining to the Creator's idea, and thereby rising to any science of the universe, and attaining what may be termed a rational cosmology. It is no presumption so to seek for this divine idea; it need have nothing of irreverence to disclose so much as may be attained; yet will it be premature doubtless for a long time to come, to announce that such idea has been completely apprehended, and may be adequately stated in any human philosophy. So much as has been gotten and given in the following pages, the careful reader will at length discover, and some may perhaps hereby be led to seek further and to see clearer. The process is directly on to the vindication of a pure Theism, and the exclusion of both Atheism and Pantheism.

The introduction to the work may seem to some to be too far extended; but as a preparative for the investigations which follow, and as an aid and a guide to the reader in the perhaps unaccustomed path he is called to travel, it is deemed that the whole will be useful, independently of the intrinsic importance it may have in itself. The first chapter may also by some be thought to have too little connection with cosmology to be here properly introduced; and yet a further attention will probably find and appreciate the advantage, before the study of the principles and laws of the cosmos, to have a carefully attained idea of a Creator as wholly independent of the cosmos he is to make and govern, beside the fact that neither Atheism nor Pantheism can ever be finally excluded except in the complete idea of an absolute Creator as distinct from Universal Nature. Still, should any find themselves both uninterested and unprofited by the discussion, they can at once pass over both the introduction and the first chapter, and commence what is properly the topic of rational Cosmology with the beginning of the second chapter.

In portions of the intuitive processes here pursued, a help might at the outset have been given to some minds by the interposition of more diagrams, and yet in the end the fastest and the pleasantest progress will be found to have been secured by casting off all dependence on any such helps, and fixing the mind's eye directly upon the subjective ideal, as the pure ground in which the insight is to attain determinations of the developed principle. In two cases only from the extent and complication of the intuition, has it seemed best to resort to the interposition of figures; in other cases care has been taken to use precise language, and to give descriptive illustrations and analogies, so that to a careful and clear inspection the process may be followed without much difficulty or discouragement. Nothing can make the journey easy to a mind that refuses to go alone and waits to be carried. The truths sought are not in the sensible phenomenon, nor at the conclusion of a logical process, but must be clear to the rational insight in their own necessity, if apprehended at all. To the intellect that does not so apprehend them, all forms of expression will be empty; to the mind that does so apprehend them, no interposed figures are needed or would be tolerated.

UNION COLLEGE, 1858.

CONTENTS.

Introduction, ... 13

Facts and principles. Facts determined by Principles. General progress of philosophical investigation. Theology and philosophy possible.

RATIONAL COSMOLOGY.

General Method, ... 55

CHAPTER I.

THE IDEA OF AN ABSOLUTE CREATOR.

1. The Absolute taken as the Infinite, ... 59
2. The Absolute taken as the Unconditioned, ... 63
3. The Absolute in the Understanding itself, ... 68
4. The Absolute as given in the Reason, ... 76

In this is found the Deity; Supernatural, Personal, and Absolute, in all His Attributes.

CHAPTER II.

THE ETERNAL PRINCIPLES OF THE UNIVERSE.

1. Matter is Force—Antagonist and Diremptive, ... 90
2. Creation—The Origination of Matter, ... 96
3. Space and Time determined, ... 103
4. Matter Perceptible by the Senses, ... 110

The Touch. Taste. Smell. Hearing. Vision.

5. Matter as Statical and Dynamical, ... 117
6. Principles of Motion, ... 120

Momentum. Virtual Velocity.

7. Creation a Nature, ... 131

8. THE MATERIAL CREATION A SPHERE, ... 134

9. THE PRINCIPLE OF GRAVITY, ... 145
Repulsion and Attraction.

10. THE PRINCIPLE OF FALLING BODIES, ... 155
Increase of Momentum. Inclined Plane.

11. THE PRINCIPLES OF MAGNETISM, ... 162
Bi-polar. Dip. Attraction and Repulsion.

12. THE PRINCIPLE OF ELECTRICITY, ... 171
Electric tension. Positive and Negative Conductors.

13. THE PRINCIPLE OF HEAT—DIREMPTIVE FORCE, ... 175
Vibration. Radiation. Absorption. Latent-heat.

14. CHEMICAL PRINCIPLES, ... 181
Combination. Equivalents. Affinities.

15. CRYSTALLINE PRINCIPLES, ... 184
Polar forces with varied axes. Geometrical solids.

16. THE PRINCIPLE OF WORLD-FORMATIONS, ... 186
Chemical Chaos. Rotating Spheres. Single and Double-Worlds. Systems. Central Suns.

17. PRINCIPLES OF PLANETARY MOTION, ... 202
Elliptical. Equal areas in equal times. Square of the periodical time as cube of the distance.

18. PRINCIPLE OF LIGHT, AND LUMINIFEROUS BODIES, ... 210
Ethereal pressure on the Sun's surface, and rotating friction. Luminous Atmosphere. Various optical phenomena.

19. THE PRINCIPLE OF GEOLOGICAL FORMATIONS, ... 218
Gravity and Cooling. Plutonian Crust and Strata. Wernerian Deposits.

20. THE PRINCIPLE OF COMETARY BODIES, ... 220
From without the system. Some caught and retained. Direct and Retrograde.

21. THE PRINCIPLE OF STELLAR DISTRIBUTION, ... 223
Hemispheral pressure and diremptive force compounded. Stellar stratification. Clusters.

LIFE.

DEMAND FOR ORGANIC BEING, ... 231

22. THE LIFE AN ASSIMILATIVE FORCE, ... 234
Formative-energy. Growth. Propagation. Sex. Species. Death.

23. THE PRINCIPLE OF VEGETATIVE LIFE, ... 240
Superficial. Ramification. Reduplication.

24. THE PRINCIPLE OF ANIMAL LIFE, ... 244
Vegetation turned inward. Muscular Irritability. Nervous Sensibility. Self-feeling.

25. THE PRINCIPLE OF HUMAN LIFE, ... 252
The Rational superinduced on the Animal. Self-centre. Self-consciousness. Supernatural. The Consummation and Crown of Nature.

CHAPTER III.

THE NECESSARY LAWS OF THE UNIVERSE.

THE CREATOR NOT THE SUBJECT OF SCIENCE, ... 256

1. THE LAW OF SPHERICITY, ... 258
Tendency in Solids. The fact in Fluids. Capillary Attraction.

2. THE LAW OF GRAVITY, ... 264
Universal, and no assumption.

3. LAWS OF MOTION, ... 269

4. LAWS OF MAGNETISM, ... 270
Magnetics and dia-magnetics. Astatic Coercive-force. Induction. Terrestrial Magnetism.

5. THE LAWS OF ELECTRICITY, ... 280
Static and dynamic conditions. Conductors. Insulators. Positive and Negative Poles. Molecular Vibration. Circularity. Electro-magnetic.

6. THE LAWS OF HEAT, ... 289
Vibrating Intensity. Diathermanous and Athermanous Bodies. Combustion. Latent Heat. Effusion. Vaporization. Animal Heat.

7. THE LAWS OF LIGHT AND LUMINIFEROUS BODIES, ... 295
Radiating Vibrations. Penumbra. Reflection. Refraction. Prismatic Spectrum. Chromatic Aberration. Interference. Polarization. Sun's spots.

8. THE LAW OF CHEMICAL FORCES, ... 307
Chemical Affinity. Definite and Indefinite Action. Simple Substances. Chemical Equivalents.

9. THE LAWS OF CRYSTALLINE FORCES, ... 314
Classification. Axial Construction. Cleavage. Contractions and Expansions.

10. THE LAWS OF THE WORLD-SYSTEMS IN THEIR ARRANGEMENT AND MOVEMENT, ... 322
Densities. Interplanetary spaces. Periodic times. Sat-

ellites. Planetary inclinations. Rotations. Uranian system apparently retrograde. Planetoids. Saturn's Rings.

11. The Law of Comets, ... 345

Cometary Origin. Incorporation into the System. Elliptical Orbits, and Hyperbolic and Parabolic Courses. Inclinations. Change of retrogradation.

12. The Laws of Geological Formation, ... 359

Broken and upturned strata. Azoic rocks. Fossil strata. Subcrystalline, Basalt, and Trap rocks. Diluvial deposits. Moon's surface, and other planets.

13. The Laws of Stellar Distribution, ... 362

The Milky-way. Stellar Clusters. Nebulæ.

14. The Laws of Life, ... 378

The Life-force Spiritual. Works to supply wants, and thus to ends. Organisms. Sex. Species. No hybrid propagation.

15. The Law of Physical Energies, ... 380

Reduced to Gravity and Heat, and thus to the two permanent and original Forces—Antagonist and Diremptive. A conversion but no annihilation of Forces.

APPENDIX.

Cosmology accords with Moses, ... 386

INTRODUCTION.

FACTS are things made—*res gestæ, facta.* They have the nature that is given to them by their Maker; and in knowing only the fact, there is no capability for knowing why their nature is thus and not otherwise. The Maker has so constituted the fact, but in our ignorance of what determined Him in the making, we can only find in experience *that* the fact is, and can by no means say *why* it is.

PRINCIPLES are truths prior to all facts, or makings, and are themselves unmade. They stand in immutable and eternal necessity; and while they condition all power, can themselves be conditioned by no power. Even Omnipotence can be wise and righteous, only as determined by immutable principles. The insight of the reason may often detect, in the fact, the principle which determined the nature of the fact, and in the light of such principle we can say *why* the fact is, and not merely *that* it is.

The perception of the sense gives facts; the insight of the reason gives principles. The use of facts may lead the mind up from particular to general judgments, whereby we may classify all the attainments of sense and secure an intelligible order of experience; the use of principles may guide the mind to interpret and explain facts, and raise its knowledge from that of a logical experience to philosophical science. Not facts alone, no matter how logically classified, but facts expounded by principles, constitute philosophy.

To know that a fact is, and to be competent to deduce a logical conclusion that because such fact is, other dependent facts must have been, or must now or in future be, is doubtless in various ways of great importance. The business and social intercourse of life could not be carried on without it. All such deductions belong to the distinct capacity of the logical understanding, and its successful cultivation secures good judgment, practical wisdom, and successful management in all economical matters. In those affairs which come within the considerations of the expedient, the prudent, the useful, such clear judgments from comprehensive facts must control, and the calculating, mercantile, business world could not get on without just such intellectual operations. The value to such operations is given from a wide experience, embracing many facts, and carefully deducing from them what other facts may be expected according to the past order of occurrences; and while one man may differ, in degree, very widely from others, yet will all men have this capacity in a measure, and their agency in practical thinking and connecting facts in general judgments will be the same in kind. Yea, a man may use more facts and conclude in broader judgments than an animal, but the man and the brute are in this doing the same work, and often the sagacity of the brute is surprisingly near to that of the human understanding.

Such well-cultivated capacity may be known as *good sense*, since it avails for the induction of many facts in sensible experience; or it may be termed *good judgment*, since it is competent to use such facts in comprehensive practical conclusions. But this is the most that can truly be said of it in its highest degrees of perfection. To call its results, in the broadest generalizations, *good philosophy*, would be wholly to mistake the name and the thing; since this practical experience can use facts only, and its most general judgments can attain facts only, while the distinctive work of philosophy is to go back of the facts, and attain and apply the principles which determine why the facts are so.

Man has the capacity for this, which the animal has not; an endowment differing utterly in kind and not merely in degree. Man can, therefore, philosophize and interpret facts, while the animal can only judge according to facts. By the insight of reason, which no animal can exercise, man attains in many facts the principle which was before the fact, and which, wholly unmade itself, controlled and guided the maker of the fact in all its construction. The objects of the most general judgments of the understanding are still only facts, things made; and if they have been intelligently made and are capable of any rational explication by their maker, or by others, they must have had their unmade principle for each, present in the mind of the maker, and that guided in his making, and which he has so put into the fact that it has become the nature of the fact, and the law of its being and working, and whose light alone can guide to any proper philosophical account of the fact it has determined. Thus, the steam engine was not as a fact, until its principle was already in the mind of its inventor, and this principle he did not make but found, and which having found, he went on to put into the fact he fabricated as the law of its peculiar being. The rational eye may readily read the law in the fact, when often the principle without the fact would not have been discovered, but when in any way the principle is attained, whether as the product of original genius or learned from his works, it is that by which we may give the explication why the fact was thus and not of some other nature. The fact not only must be known, the principle which was before its making must also be known, or we can have no rational philosophy about it.

Now, just such application of eternal and immutable principle is demanded for the philosophical study of universal nature. Observation may give its many particular facts, and general conclusions from broad inductions may assume to have found facts of universal comprehension, yet are these highest facts necessarily, thus, unexplained facts, and as without any known principle themselves, they must be ever wholly incompetent to lead to any philosophical interpretation of the included facts which may be classified under them. One fact may thus be gained, as that which shall make all facts *turn together* in it, and thereby we may have literally a *universe*, still we can thus have it and its included universe only as a fact, with no possible rational philosophy of any thing. If we know the fact that nature is a universe, we have no principle by which we can at all interpret why it is so.

Thus, by wide experiment and profound calculation the great fact of universal gravitation in matter has been assumed, and the conclusion has been reached that all matter gravitates toward all other matter, directly as the quantity and inversely as the square of the distance, and we bind nature in a universe by it; but at the most, this is only given as a fact, with no principle that has so determined it, and it can therefore only give the universe as a fact and afford no possible rational explication of it. If we have not the unmade principle determining the fact of gravity so to be, and with just such ratios, then have we no rational science of nature, and what we call a law of nature is still a bare fact; an arbitrary making; and no philosophy interpreting the making by its principle. The vast superstructure we have reared is all the work of the logical understanding, without one ray of the expounding reason to shine on it and through it. The whole frame-work has been put together, with much of human toil and din, from the outside, but no eye has found and fixed its absorbing gaze upon that inner force which, in the reality, has been silently making living stones grow together to be the Lord's holy Temple. Till we attain this eternal principle, which as a living law the Maker of the universe has diffused all through it from centre to circumference, we may stand on the outside and measure and weigh, and overwhelm the understanding with the summations of arithmetical reckonings, but we shall know nothing of that central working which makes and holds all in one concrete *cosmos* of perpetual harmony and beauty.

Universal nature is more than bare fact; it is something made under the determining conditions of unmade principle: and this immutable principle, under which its being and all its ongoings have been determined, has now its counterpart in nature as the perpetual law of its working, and the human reason may find at least some glimpses of it and interpret the great plan by it, and may so far know what nature is, and why it is thus, and not forever rest in the mere knowledge that it is. If, indeed, we cannot extend our knowledge beyond the bare facts of experience, then must we perforce content ourselves with the mere phenomena of nature, but we may not assume that any such knowledge is a *science* of nature, for this cannot be attained except as we reach and apply the determining principle. A rational cosmology is the only true natural philosophy.

This immutable principle, which determines how the fact may be, and, if the fact be at all, how it must be, is given in pure thought alone, and can be no appearance in the sense. Neither can it be that which connects the qualities given in the sense into one thing, for that is effected in the substance; nor that which connects the successive events into one series, for that must be done through the cause; but the principle lies still further back, and determines the natures of substances and causes themselves, and stands as the archetype or ideal pattern after which the essential natures of things have been created. It is the consistent thought, as idea, how the fact may be, and when carried in combination through all facts, it becomes the consistent idea of how a universe may be. All the statics and dynamics of nature were arranged by it, and thus it was before the forces of nature and their balanced action became facts, and therefore existed as a subjective ideal in the mind of the Maker of the universe only. The principle as in being before the fact, and which is to determine the fact, has not yet been brought out into objective existence, but subsists as mental being alone. The principle, thus, is not science, but only the ideal of a possible being, which, when it shall become fact, may be subjected to science.

As a general illustration of the being and application of all immutable principles, I may adduce the subjective thought of an arch, or of a catenary curve, and may so apply these in a completed projection as to have the ideal of a standing or of a hanging bridge; or, I may take the subjective thought of a mechanical power, and follow out the composition and resolution of forces till I have projected some ideal engine; and I shall then have the bridge or the engine in pure thought, and which will be subjective pattern of what the bridge or the engine must be, if they become manifest in objective fact. At the most, here will be the science of the possible only. It may even be that nature will not admit of these ideals becoming facts. Perhaps my projected structure is such, that no actual materials would bear their weight in the bridge, or the pressure of such a force in the machine, and then my perfectly consistent thought could never be made an actual thing. The theory is self-consistent, but the fact would be self-contradictory. The science cannot be complete until both the principle, as self-consistent thought, has been obtained, and this principle has also found its actual counterpart, as the existing law of the combined materials.

So, on the other hand, I may have seen a bridge resting on such a material arch, or suspended on such chains; or, I may have seen a machine moving with such a mechanical power; and then I can, by experimental measure and weight, make other constructions like to these, and thus actually put the law of the models into the facts I have made to imitate them. But inasmuch as my work has been only an imitation, and I have recognized and applied no law in the facts which had been determined by an immutable principle, I cannot be said in having the fact, to have any proper science of it.

Thus the subjective idea alone is not complete science; and the fact as mere fact is not science; the first is only the knowledge of the possible, the last is only the knowledge of the empirical; but when the subjective idea as the principle determining the fact, and the objective law as put by the principle into the fact, are both attained as accordant counterparts of each other, we have then both an interpreting principle and an interpreted fact, and in this is complete science. The whole process in its attainment is a rational philosophy. A Rational Cosmology must conform to this criterion of all science, and only in so far forth as it is kept within the constant circumscription of such criterion can it have any claim to a rational philosophy. All that is fact—the entire cosmos, as a making after a principle—may be so subjected to philosophy by an adequate insight of reason.

But the cosmos, or world of fact, must have its Maker. A universe, coming up successively or collectively out of a void of all being, would be an impossible conception. It would oblige the understanding to think a substance that was not substantial, and therein to think an absurdity. This Creator of the cosmos must be wholly absolved from all the conditions determining the cosmos; he must originate it, and give to it its nature while he is wholly supernatural; and thus, as the author of all fact and not himself a fact, or a making, he cannot be subjected to any science by the finite reason. It may be demonstrated that God exists, and that he is absolute, in the sense of complete absolution from all the conditioned necessities in nature; but there can be neither a principle as archetype after which he was made, nor a law which works in him as a constituted fact, and subjecting him to its nature, and thus the criterion of all science is inapplicable to the Deity as subject to philosophy. When we have demonstrated that God is, and that he is absolutely supernatural, we have all that Theology demands, and do not need to bring him within the definitions of philosophy. From the nature of the case philosophy must recognize theology; neither can exclude the other, nor can the one be identified in the other. There is a dualism; the world is not without its Maker, and the Maker is not in and of the world; the theology rests on the proof *that* God is, the philosophy rests in the interpreting *how* the world is; and all philosophy without theology is incomprehensible, and all theology without philosophy is a credulous superstition. All blending and confounding of the two will be destructive of both. If the universe be absorbed in the Deity, it is Pantheism; if the Deity be lost in the universe, it is Pancosmism. But the unphilosophical Pantheism will be Atheism, and the atheistic Pancosmism will annihilate all philosophy in absurdity.

The whole design includes the attainment of a clear conception of what is essential in a Being that must be the Maker of the universe; and then, a clear conception also of the immutable principles that must determine the laws, and by which we may expound the nature of the universe. The Maker must be an absolute personal God, capable of originating material worlds from himself, without himself being subjected to any of the conditions of matter. But we may rest in the demonstration that such a supernatural Being is, without attempting the solecism of attaining a principle that is philosophically to interpret the absolute *principium*, and determine why he is. In reference to the theology, there may be complete satisfaction attained in the use of a true rational Psychology; but the new and severe task demanded is in reference to the philosophy. There is the necessity for the instauration of *a true science of the universe*—A RATIONAL COSMOLOGY.

It will assist much in setting clearly before the mind the urgent necessity for such a work, if we rapidly look over the track of past philosophical investigation, and notice the prominent attitudes in which philosophy has stood, and the positions now occupied by distinguished schools or the representative men who speak authoritatively for them.

In the earliest ages of Grecian history we find the dawn of all philosophical thinking, so far as any light has come down to our day. This thinking consisted in the construction of theories, more or less crude, concerning the origin of material nature and the arrangement of the world. The various early cosmogonies, though partially and obscurely transmitted to us, are sufficient to determine what was the scope and bearing of their philosophical speculations.

The germ of any intelligible theory is first found in the recognition of some of the elemental forces in nature, and assuming that their action was sufficient to account for the formation of the universe. The natures and powers of these elements were taken as already in being, and each philosopher assumed and applied them in speculation, as he deemed them to be the most favorable in accounting for the varied phenomena. The Ionic class of philosophers were among the earliest, and their philosophizing was mainly in the above method. Thales made the element of water to be the chief ingredient in the composition of material nature, and taught that the forces here acting had been the primitive agents in the construction of the universe. Anaximenes, in a similar way of applying the elemental forces, held that the air had given the first formative processes in the arrangements of nature; and Anaximander had some vague conception of higher elemental powers not in any distinct form of manifestation, but existing as a chaos of rudimental being, out of which an orderly arrangement ultimately emerged. These recognized, each in his way, the presence of efficient agencies already in existence, but seem not to have arisen to any speculative conclusions concerning the origin of any of these elemental forces that they assumed as active in the formation of worlds. There was some first cause, but they did not go beyond already existing elementary forces to find it.

Pythagoras is one of the most conspicuous of the early philosophers, and enough is transmitted to us to prove that his clearness and force of philosophic thought was quite beyond the age in which he flourished. He seems to have apprehended the distinct faculty of the human mind to attain to truths beyond the sensuous perceptions, and to reach necessary and immutable principles. The axioms which determine in the combinations of numbers, and the regulative proportions in mathematical formulæ, and the harmony ot tones in music had been intuitively apprehended by him, and he had hence learned to guide his philosophical speculations by those permanent truths that must condition and correct all the fleeting perceptions of the sense, and by which must be interpreted and explained all the seeming anomalies and contradictions in the phenomenal world. He had learned to apply principles to facts, and thus had found the right method for a true philosophy. The effort to clothe his systematic thought in mathematical phraseology and to represent the physical forces of nature under the forms and ratios of number, has left very much that remains to us of his philosophy, from the representations of those who followed him, quite ambiguous and obscure; but it is still easy to gain a correct and profound meaning from many of these representations. Others, that are so enigmatical that little can be made from them, were probably clear in his own apprehension, and need now only the necessary clue to lead us through the obscurities to a consistent meaning.

The origin of the chaotic elements of the universe was not yet approached in their philosophizing, nor had there been any distinct conception of some independent author by whom a proper creation, a beginning of things, could be made. Parmenides argued that non-being was inconceivable, and that as something could not come from nothing, therefore creation, in the sense of absolute origination, was impossible. Empedocles also taught, that the elementary matter of the universe, in the *hyle*, or primary rudimental substance, was itself uncreated and indestructible. Heraclitus taught that this elementary matter was in constant flux, and that such perpetual flow of the component elements kept nature in a continual succession of becoming and departing phenomena; but he recognized nothing that could originate and orderly control and guide these flowing movements. He held all things to be of fire, and yet not in the same way that the Ionic philosophers had applied the forces of the natural elements, but rather because fire is of so penetrating a nature, and decomposing other substances, and thus keeping nature perpetually fluid and agitated. Later, among the Sophists, Protagoras took this constant arising and departing, as the necessary result from our mode of knowing, and in which all things must be fleeting and transient as our sensations present them. Man was made the measure of all things, and to every man, his own consciousness in his perceptions must be to him the truth. What his senses gave, that, to him, the things themselves were, and every man must follow his own measure.

The old atomic philosophy, again, reduced all of nature to an original being in indivisible and indestructible atoms, and brought those atoms together in bodies, either by a falling together, or by an inner deflecting force, which turned them out of their proper course in their descent, and thus collected them in masses. There was no occasion for a creating and superintending Deity, for all things were provided in the original atoms. The whole philosophy was entirely atheistic.

Anaxagoras seems first to have found and traced the indices of some intelligent adaptations to ends in nature, and that such adaptations were the evidence of design; and he accordingly taught that there was a Mind concerned in the formation of the worlds from their chaotic state. But this *νοῦς*, or intelligence, was apprehended rather as subjectively in the world itself, and a kind of inworking power that ordered and arranged its changes as an indwelling law, than any independent and personal agent. With him, there was no rising above nature and apprehending a supernatural and rational Creator and Governor, but merely an attainment of the facts of design, and workings of an inward intelligence, without referring them to any thing beyond nature itself. The world was, and had its own intelligent activity within itself, and thus the universe was mind as well as matter.

Plato was the great master philosopher of the age. He not only recognized clearly the *νοῦς*, or intelligence, manifested in the adaptations of nature, as they had been found and taught by Anaxagoras, but he referred this intelligent adaptation to ends, directly to a supreme Deity. He apprehended also, more clearly and comprehensively than Pythagoras, those necessary and immutable principles, which, antecedently to all facts, regulate and determine in the production of facts, and necessitate the conditions in the ongoings of nature. He is emphatically the rational thinker of humanity, and his conception of philosophy that which must correct all subsequent erroneous methods of speculation. Only in returning to his method, can modern wanderers in new paths be turned about again into the old and safe highway. With Plato, the universe stands out as one consistent whole in itself, and this universe the product of an independent and personal Creator. The Absolute Good had, from eternity, the Ideas, or Archetypes, in himself, and he produced and fashioned the universe from himself accordingly. Xenophanes had, before this, generalized the many into the one, and made all to stand as parts of the whole, and had called this whole, God. He was in truth, the first philosophical Pantheist. But Plato's whole was the whole of nature only—the created universe—having the Deity utterly above and independent of itself. Sometimes, it is true, that the insoluble difficulty of accounting for evil under the absolute dominion of the Good, leads Plato to reason as if matter was the source of all evil, and that this had an existence, as it were, independent of God, and in this way freeing God from connection with evil which, in the necessity of the case, could not be excluded. But this is not the doctrine systematically held and taught by Plato. In the Timæus, the matured and labored philosophy of Plato is given; and here we have one supreme Absolute Mind, producing the Universe from himself and making it one living whole by infusing all through it the *informing* Idea as the soul of the world. God is, and then the world is made by him, and the intelligent Idea or law is put into it, and thus nature moves on, as a living thing, to fulfil its grand design.

The Platonic philosophy has its *first mover*, in the acceptation of an uncaused originator. Movement is not only locomotion, or progress in space, but it includes all changes. Motion in space; growth and decrease; arising and vanishing; beginning and annihilation; the inner activity of thought and all spiritual agency; all involve the conception of movement; change; and necessarily imply a constant or permanent, from which all change must spring That which is mutable, and thus perishable, has been generated from that which is unchangeable and eternal. The mutable is the subject of sensuous knowledge and comes within experience; the constant and eternal can be cog nized only in the rational intellect. An immutable and eternal God, having in himself the patterns, or perfect ideas of all things, generated the Universe from himself; vitalized or ensouled it, by putting the eternal Ideas into it; thus making nature to possess a living force and an orderly intelligent activity. The Universe is itself, thus, a true good, as the free product of the absolute Good; and having efficiency, activity, orderly intelligent progress, it is spoken of by Plato as if it were itself a living thing, "*a blessed god.*"

This Platonic philosophy completely avoids both Atheism and Pantheism, and is thoroughly Theistic. The pagan polytheism which it recognizes is in no sense contradictory to pure monotheism. The Absolute Spirit is ever held as supreme, independent, and eternal. He first makes soul, as better and thus older than body; and from this soul of the universe, as originated direct from the Absolute Good, there is successively generated all other spirits, and with Plato, all spiritual being is a god. The Absolute Good is, however, with him the God of all gods. The philosophy falters in nothing that is necessary to a true personal Deity; a God utterly supernatural, and wholly distinct from and independent of the universe which he makes and governs. The theology is conceived and preserved pure and unadulterated from any material conditioning or physical necessitating. But while his philosophy of the material Universe proceeds always in the true method of accounting for fact by principle, yet is there not unfrequently a very imperfect apprehension of principle, and thus often a wide misapplication of it. Physical facts were but partially attained and confusedly apprehended, and the age of humanity was not then sufficiently advanced to be able to read clearly the law in the fact, because of this imperfect comprehension of the fact. The insight of reason was, with Plato, superlatively penetrating, but the ground in which the eternal principles must reveal themselves was not plain and full before him. The great fact of a creation was clear, and he saw in it the certainty of a free and independent Creator; and the great truth, that this creation must conform to the immutable Ideas, or principles of absolute reason, was clear, but all these principles could not be exactly attained, because the laws in the phenomenal facts which disclose them had not been minutely observed. Only reason can see the principle in the fact, but to reason, the apprehended fact is often the only ground in which the Eternal Idea will present itself. The creating genius, which may originate its own subjective conceptions, in which it shall beforehand see all objective laws that shall exist, would be more than human.

The great merit of Plato, therefore, is not the fulness and exactness of a religious system—for in many doctrines there is the deficiency and error which was to have been anticipated in a pagan—nor the thoroughness and faultlessness of his system of natural science—for his ignorance of many facts made him falter in the attainment and application of many principles—but the prompt introduction and steadfast maintenance of the true method of all philosophizing relatively to the origin of the universe. His conception of a true rational cosmology is perfect. He has both a theology and a philosophy, and he puts and keeps both in their proper places. He never degrades the supreme Good to be the mere *animus mundi*, nor does he exalt nature to the throne of the Deity. His "soul of the world" wholly dispenses with the necessity of a *Deus ex machina*, and gives to the Universe perpetual efficiency and movement; but this infused intelligence and power is still the creature of God, and working orderly and rationally. Plato never contents himself with bare facts, but the fact is as nothing to him till he can bring it under the determination of a principle. He recognizes in the supreme Good, an agency that can absolutely begin; an independent personality that can originate from himself, without the supposition of an already previously constituted nature causing him to do this. His God is supernatural; Spirit in liberty; Absolute personality; and the created Universe is the free rational product of this God; intelligible and wholly explicable from the eternal Ideas; a consistent cosmos; fact pervaded by principle.

Since the age of Plato, philosophy has been little Platonic. The New Academy had nothing of his spirit. The New-Platonism of the Alexandrian school was, also, altogether a corruption. The blending of Orientalism made it a perversion and not a perpetuation of Platonism. The intellectual vision, by which the human soul apprehended the eternal Ideas, and came to the recognition of the supreme Good, was turned to an absorbing silent meditation, in which it was sought to identify the contemplating philosopher with the contemplated Deity, and give the human soul to be swallowed up in the divine.

The ARISTOTELIAN PHILOSOPHY at once, after Plato, struck determinately into quite another path. The too much extended and thereby confused application of the Idea, by Plato, had made the Aristotelian philosophizing necessary. In the Idea Plato had included not only the archetypes which were eternally in the Divine Reason, and the primitive forces which are the principles or germs from which universal nature is developed, but also all general conceptions from which, by virtue of their participation therein, all the particulars of the class have their being, and these ideas in this broad sense were also held by him to be true and valid realities; it thus became a demand of the reason that this broad assumption of real being should be critically examined. Originally, in its founder, the Aristotelian philosophy used the insight of reason, and recognized the Eternal principles necessary for all facts, as really as the Platonic. All physics was made to strike its root and find its explication only in metaphysics. The *prima philosophia* was essential to all philosophy. But the study was intently and intentionally turned to follow out nature on the phenomenal side, and not to hold philosophy perpetually under the control of eternal principles. Generalized facts were themselves put as principles, and a classification of phenomena under general facts came to be recognized as philosophy. Genera and species, put as categories under which might be classified all particulars, took the place of the eternal ideas, and instead of recognizing any being above sense, the veritable and immutable ideas of Plato were said to be only "things of sense immortalized." At length, among his later followers, empty words, names instead of things, absorbed the whole attention; and the purely logical understanding became the entire faculty for philosophizing, and this wholly exhausted itself in running through all the processes of syllogistic reasoning. Reason, as "the vision and the faculty divine," distinct from the faculty connecting in logical judgments, was so completely disused and overlooked, that it ceased to be recognized as a distinct fact in psychology. The law of the syllogism admitted no distribution in the conclusion which had not already been gathered in the major proposition, and the whole labor only analyzed what they had, but added nothing new. Experience attained all the facts; abstraction and generalization gave the logical notions; and the syllogistic process analyzed and distributed in specific conclusions. A clearer knowledge of what they already had was secured, but nothing new was added, and nothing philosophically expounded in its principle.

On emerging from the long and unsatisfactory strife of the scholastic logic, the human thought turned mainly into two distinct channels. CARTESIANISM, having some alliance with Platonism, ran out its course the earliest. This philosophy awakes in doubt, and casting around for what may resolve all doubt into clear certainty, and assuming that clearness is the test of truth, it finds an undoubted fact of thinking clearly in the consciousness. Here is the starting point for all philosophy, and hence the famous dictum of Des Cartes—*Cogito, ergo sum.* This is as much as saying —there is a thinking, and by thinking myself is found. Extension is also as clearly given in the consciousness as thought, and these two, thought and extension, are the essence of all being. The first is distinctive of spiritual being, and the last of material; and these two are so wholly unlike and disparate, that no intercommunion can subsist between them. All interchange of activity between mind and matter, must be effected by the interposition of the Deity; and hence the general doctrine of "divine assistance," for all communication of the material with the spiritual. The Deity was an *à priori* assumption, from the prominence and clearness of the idea, which in itself involved a necessity. Extension, with its two modes of rest and motion, admitted of being broken into parts, and hence the atoms; hence, also, the vortices induced in the universal breaking up, the collection of the differently ground up atoms into their appropriate spheres, and thereby the general arrangement of the universe.

Geulincx added the perpetual interposition of the Deity, in all occasions when the spirit acted upon matter, or was affected by matter, and thus introduced the doctrine of "Occasional Causes;" and Malebranche reconciled the spiritual perception of material objects, by the existence of all things in the Deity; and thus, through this divine medium, matter could be perceived by spirit, and hence his doctrine that "we see all things in God." Spinoza ultimately finished this order of thought, by bringing the duality of thought and extension into complete unity, and identified both in a higher Infinite substance. This Infinite substance is made the ground of all being; and all the various manifestations of both thought and extension, spirit and matter, are but the varied attributes of the one Infinite substance.

Leibnitz, it is true, changed the dead atoms of Des Cartes into reflecting or envisaging monads, and pre-arranged them so as to give their representations harmoniously one with another, and made this "pre-established harmony" to fulfil the purposes of the "divine assistance" and the "occasional causes" before given; but the Cartesian philosophy is truly consummated in Spinozism. No movement of thought can pass beyond the Infinite substance; and all theology, and all philosophy, have the same source, for the Infinite substance is the only God, and the philosophy of the universe is but the recognition of God's manifested attributes. The Infinite substance, when subjected to reflection, is truly only a *substratum* for the phenomena of thought and extension, and is itself wholly dead and inert, except that it admits of these attributes to inhere in it. As a theology, it could not satisfy; for this dead, inert, impersonal substance, was nothing that could be loved or worshipped. It solely sustained the spiritual and material worlds, but it could neither create nor govern them. As a philosophy, it could just as little satisfy; since, although it furnished a unity for the disparate conceptions of thought and extension, yet was it a mere logical unity, and, though placed at the centre, could exert no efficiency and possess no intelligent law or rational principle. The philosophic thought, dwelling upon this Infinite substance, could do nothing with it, nor make any thing out of it. It revealed nothing, it interpreted nothing.

Cartesianism began with the Platonic views of a "first mover," and the competency to attain and apply *à priori* principles; and the philosophy was carried onwards by attempting to apply the insight of reason, and follow the determinations of eternal ideas. But this was made absurd and impossible, since matter was essentially mere extension, passive, inert, and lawless; and even all assumed spiritual divine action upon it was in violation of its fundamental doctrine—the essential incommunicability of all spirit with matter. At last, both spirit and matter were put in a substance which merely held them in identity, but could neither use nor control them. God and the universe were one; but the pantheistic unity was utterly dry and dead at the heart, for it had no personality there with which piety could commune, and no principle there with which philosophy could work.

The other channel of thought was the BACONIAN INDUCTIVE LOGIC. This has run a much longer course, and turned in more varied directions, and yet it has mostly kept itself at a further remove from the great Platonic requisition, that all philosophy must maintain an interpreting immutable principle at the centre.

All the old scholastic syllogisms were built upon the analytical dictum, that what is true of the whole must be true of all the parts. This could lead to no extension of knowledge, for it obliged that the truth for the whole should be attained before it went to the work of distributing to the parts. The Inductive Logic exactly reverses the dictum, and builds upon the judgment, that what is true of all the parts is true of the whole. This allows scope for extending knowledge, for it encourages and obliges to the attainment of the truth for all the parts before concluding upon the truth for the whole. We might anticipate that such an impulse would not rest in barren results. All means to attain the truth of the parts will be desirable, and at once put in requisition. If, now, we may here rest upon the Platonic method, and from the insight of reason can affirm that nature, as itself a fact, has been made after the determinations of eternal principles, then we know that such determining principles must run their lines all through nature, and we shall find no fact in nature that is not bound up by laws with its fellows. Instead, then, of trying to attain all the facts which go to make up the whole by a particular experiment for each, we may be safely content with an experience that reaches so far as fairly to convince that nature's law has therein been found, and then we may cease from all further experiment, and logically conclude upon the truth for the whole. We have found nature's law, and we know that this law must hold all the parts, and the short turn of logic answers for all the long labor of a universal experience.

So Bacon, and long before Bacon, so Aristotle philosophized; and hence the *organon* of the latter, and the *novum organon* of the former. But if the Platonic doctrine is all assumption, and the reason's insight of principle and law in nature is a delusion, then must our actual experience run through every part before we may at all conclude upon the truth for the whole. The inductive logic is open to skepticism on all sides, so soon as we deny that reason is capable to attain and put eternal principles at its foundation. Without this, we have no right to assign any laws to nature, and can only say, so far forth as experience has gone, so the facts are; but we have nothing to sustain our footsteps beyond experience. And if we should deduce a general judgment from an induction of many particulars, as if the actual experiment had extended to all, such an assumed general judgment could only include the bare fact, and could give no principle that could interpret it, and we could only use it to classify particular facts under it, but not in any way philosophically to explain them.

Thus, without the insight of reason, the inductive logic begins and prosecutes its work in credulity, and when it deduces its general fact, it can never evince its validity, and the assumption can only be of a dry, hard, insoluble fact, which can never find its principles to explain why it must have been thus and not otherwise. We may make one fact dependent upon another, and thus upward through an indefinite series, but we can reach to no principle that supports and expounds the whole chain.

A true inductive process must both begin in the application of immutable principle for the determination of nature, and its most general facts must themselves be interpreted by principles which determine them, and then it becomes a safe guide and auxiliary to philosophy; but when used by such as discard the insight of reason, and deny the power of the human mind to go beyond the fact, it becomes not merely useless to philosophy, but is itself utterly unphilosophical. At the best, it cannot itself be a philosophy, but only an instrument in the interest of philosophy; but as now mostly used, in the rejection of all *à priori* principle, it is wholly illogical and illegitimate. In its own proper field of attaining facts in the service of philosophy, and thus for enlarging the field of discovery, the inductive logic has done much, and become the wonder and boast of the age, which, as practically utilitarian, has been fashioned and almost wholly actuated by it. Let it have its due, but let it not usurp honors which are not its due. Let it be employed to the utmost in its proper field, but let it not come out of its place, arrogantly to dictate in matters about which it can know nothing. It must perpetually walk in the borrowed light of a higher faculty, or it becomes inevitably both unphilosophical and atheistic.

Take the inductive logic alone, and cut off all communication of immutable principles in the insight of the reason, and proudly as she may seem to walk over the field of phenomenal nature, yet can she vindicate her possession logically to no fact she assumes beyond actual experiment, and can never expound a single fact she gathers, nor ever cast a glance within the region of the supernatural and eternal. Make this the highest operation of the human mind, and absolutely shut out of human possession all knowledge that it cannot attain and vindicate, and a personal, absolute Deity can then be neither proved nor conceived, and you thus first exclude, what must then be, the gross delusions and credulities of theology. All facts are then also mere facts, with no eternally conditioning principles to determine them, and you thus exclude, what must then be, the illusory and bewildering lights of metaphysic. The theologic age, old in its venerable but mischievous superstitions, passes utterly away from the generations of humanity; and next, the metaphysic age passes, with its lofty and profound, but empty speculations, neither of them again ever to return. All religion and all philosophy have passed beneath the horizon, and the complete and final positivism of Auguste Comte culminates in the heavens. In this the full mission of the inductive logic is accomplished. She began by denying to the human mind any higher light than experience; she carried out her varied experiments, and brought together numerous kindred facts, and deduced more general facts from these conspiring individuals; she arranges all carefully according to variety and class, species and genus, and with her light shining fairly but exclusively upon these arranged facts of phenomenal nature, she finds the bold man who does not shrink from her logic, and who well knows that no modern speculative school can rebuke him, and he cries aloud to the nations of the earth—All theology and all philosophy beyond this is a fable.

Positivism is the affirmative side of Hume's skepticism, and rests firmly and impregnably on the basis of the exclusively sensational psychology. All the elements of possible human knowledge are affirmed to be given in the senses. The understanding can reflect upon these, and abstract, compare, and combine, and thus attain new analytical judgments out of them, but it can add nothing more, and attain nothing other than is given in them. What is made, what comes as event, we can know; but the principle determining the making, and the order of the coming, we cannot know; all expectation that nature will go on in future, in the order of sequence as in the past, rests solely on the experience having become accustomed to it. Science can affirm nothing about it; for that there are any principles beyond the facts, which have put their determining laws within the facts, is beyond all human ken not only, but all human conception. The principle must itself have been once made, and even its very maker must have had a constitutional nature that might have been directly reversed. It is philosophical to doubt, in every case that cannot become a fact for the senses. Thus Hume; and Comte is exactly the counterpart. We can positively affirm so far as experiment testifies; we can as positively deny all conclusiveness to any affirmation not capable of the testimony of experiment; facts, and facts only, are positive. And now, these conclusions are all logically inevitable from the premises. The psychology cannot be retained, and the immutable principles of theology and philosophy be admitted. If the insight of reason is not something other and higher than any judgments of the logical understanding, whether deductive or inductive, and if the human mind cannot vindicate its right to the possession and application of these immutable principles, then Hume has the right to doubt, and Comte the right positively to deny, that man can have any stable theology or philosophy.

An attempt to escape from this rigid exclusion of all stable theology and philosophy, is vainly made by that which calls itself THE PHILOSOPHY OF COMMON SENSE. This rests on the fact it finds, that the human mind is forced to assent to what are called "first truths," or "primitive beliefs," and assumes that in these there is a sufficient basis for theology and philosophy. Its strong ground is, that from the constitution of the human mind it cannot expel the convictions occasioned by these "first truths." The skeptic and the assumed infidel are forced to the same conviction, and can never belie this coercion of common sense, and can pretend to be free from it only in their speculations. They must rest on some primitive conviction, or all affirmation of doubting would itself be absurd. Neither skepticism nor positivism could affirm themselves, except by admitting the conclusiveness of common sense.

The *argumentum ad hominem*, so pushed, may seem to silence the gainsaying skeptic, and confirm the credulous disciple, but it is wholly sophistical and delusive. Common sense begins in the affirmation of the same dictum with Hume's Skepticism and Comte's Positivism, viz.: that the human mind can never carry its knowledge beyond facts. But it seeks to escape the rigid logic of the unbeliever by affirming that it finds this deeper fact in humanity, that all men must yield assent to the force of their "primitive beliefs." True to its fundamental dictum, *nothing but facts;* it makes this conviction of common sense to be mere fact, unavoidable, but yet wholly inexplicable. The human mind is so made.

The old Grecian κοιναὶ ἔννοιαι—the *common rational intelligence;* the endowment which distinguishes the man from the brute—is held to be a mere fact, and the affirmations of reason to be as arbitrary a making as the constitution of organic sensation, and thus there is felt no scruple in translating this term which expresses man's highest prerogative, by the utterly inadequate expression, *common sense.* The sum is this—the human mind is so made, that there comes out the universal fact, of a necessary assent to the "primitive beliefs." All is an arbitrary making; unintelligible, insoluble fact; and nothing unmade can be reached that may give any explanation.

And now, what does all this, but wrap the same strong chain of Positivism one fold more around our human knowledge, and make its bondage to sensationalism, and its exclusion of all theology and philosophy the more hopeless? Common sense is a thing made; and its primitive beliefs are things made; all unmade principle is beyond knowledge or conception; and even the Deity can come only within this common sense conception, and himself, and his principles of working and governing, and the whole supernatural field of immortality, must fall back within the sphere of constitutional existence, for all truth absolved from the conditions of a nature of things is wholly inconceivable. The Creator who makes worlds, and the mill which grinds corn, have alike their constitutional adaptations to their work, and our conceptions of them can differ nothing in kind, only the one has a constitutional nature more magnificent than the other! When the supplied common sense is itself only fact, and its highest attainments are but facts, then surely common sense should admit that its theology and philosophy can deal with nothing beyond facts.

The deficiencies of sensationalism, and their logical consequence in skepticism, gave rise to the CRITICAL PHILOSOPHY. In many respects, this is one of the most remarkable, and in some respects, the most productive direction in which the stream of human speculation has been turned. Kant saw the inevitable skeptical issues of Empiricism, and hence his *Critik of Pure Reason*, to escape therefrom. His method is wholly Aristotelian, though he gathers his facts in another field, and not at all Platonic, although using some of Plato's terms. He does not start from immutable principles, in the eternal Ideas, and determine therefrom how *all* judgments in an understanding must be, but he takes our human faculty of judgment as already made, and by a transcendental analysis of it determines how *we* must know. It is a critik of *pure reason*, in the sense of taking the facts of human psychology antecedently to their development in phenomena, but not antecedently to their being in subjective faculty; before they come out in our consciousness, but not before they have been constituted. The whole philosophy is *à priori*, or transcendental, not as attaining principles prior to, or transcending facts, but only as attaining facts that exist prior to, or transcending, our conscious experience.

The Platonic reason attained and used the eternal, unmade principles, or Ideas; the Kantian reason attains and uses the regulative forms in an already made human understanding. This truly Aristotelian method prevails in all the successors of Kant, in carrying forward the critical philosophy; and the pure thinking is no insight of reason that gets in the facts their determining principles, but solely an analytical process that finds facts already in the human mind before they have worked themselves out on the field of consciousness. The whole labor, though transcending the point of conscious experience, is still that of the logical understanding only.

Kant assumed that the organic content, given as envisaged by the sense, was real; but that the human mind possessed its own forms, or regulative conceptions, and these gave their law to the operation of the mind in knowing this content in sensation. To *us*, so made, our cognition of objects must therefore conform to these inherent regulative conceptions, or categories, in our human understanding. However the things may be in themselves, or however other minds may know them, our human knowledge must be after these forms already existing as facts within us. The *matter* of our cognitions had, thus, an objective reality, but the *forms*, in which our understandings clothed the objects, had only subjective reality. Kant could thus answer Hume,—we connect the sequences of events in nature in the conviction of a fixed series of cause and effect, not because our experience has become accustomed to such an order, but because such is the law of connecting in judgments by the original constitution of the human mind. But this has still subjective certainty only.

Our minds must know through the connections of cause and effect, and the other categories given constitutionally within them; perhaps other minds may know the same things in quite other connections. The universe, and the Maker of the universe, can be cognized only through these regulative conceptions, and as we can have no phenomenal content in sensation of the Deity, so we cannot demonstrate his existence, but also just as little can we carry our demonstration against his existence. The proof for any supersensible existence is from the *practical* and not from the *speculative* reason. The fact that we are thus constituted, having by the Critik been transcendentally found, enables us to say *a priori* how far the human mind can know.

Fichte, pursuing the transcendental critik still further, showed that there was no more ground for holding the organic content, or matter, for our cognitions, to be objectively real, than for holding the forms to be so, under which our understandings brought it. Both are subjective, and the matter and the form are alike supplied for the consciousness by the working of the intellectual self, or the ego. The self, as subject, makes itself object to itself. The mind can envisage nothing that it does not itself set before it.

Transcending consciousness further than Fichte, and going deeper into the mysteries of human cognition, Schelling, by an "Intellectual Intuition," detected the absolute ego standing in the mid-point of indifference to either subject or object, and as a bi-polar agency, like magnetism, simultaneously working each way, and on the one side giving the object, and on the other the subject.

He identified both the subject and object in this central ego, which, back of consciousness, works out its two poles into consciousness, and there they appear as separately Object known, and Subject knowing.

But even beyond this analysis, there was still an insoluble element lying in this absolute ego. As a source for both of Fichte's subjective and objective egos, Schelling had placed in the mid-point of indifference another ego, and which, as the absolute ego, gave both the others to the consciousness with one undivided act. This bi-polar agency, in the absolute ego, Schelling had assumed without any examination or explanation, and with such a thought-agency assumed, he could work out the process of its development into universal nature, humanity, and completed Deity, with great precision and exactness.

Hegel took this unsolved agency of Schelling, and carried the transcendental analysis to a still deeper abstraction, for the starting-point of another philosophical development. In his Phenomenology, he sets out from the common conviction that there is a dualism of both subject and object in human cognition, and thence unweaving the dialectical web in which both had been gathered, he found, as the ultimate remnant, a simple thought-progress—a movement according to the law of thinking; a pure activity with no ground—and from this abstract thought-process, with no substantial ego, Hegel begins his philosophy, and evolves the objective universe, the subjective mind, and finally the universal mind, educated to self-consciousness, and also to the knowledge that the thought-process is the only reality.

The phenomenology is solely preparatory to the philosophy, which must begin in this pure thought-process. Instead of standing at the outside and looking on, as Schelling had done, Hegel puts himself within the thought-movement. This is a peculiarity to be marked. The student of this philosophy must not at all look on, nor look forward to forecast what may come, but must absorb his attention in the movement itself, and let the process bring out in its development what it may. In this method, he is made to think over again the great thought of the universe.

The critical philosophy is consummated in Hegelianism. No passage can be opened to any further speculation in this direction. The philosophic life in Germany is in suspended activity, and must so be retained until the apprehension of the incompleteness of the critical method shall induce to the setting of some new germ in quite another soil. It began in the attainment of facts which transcend consciousness, and from these determined the modifications that must be given in consciousness. From Kant to Hegel it successively threw off more and more of that which had any objective reality, till it found itself at last with only a thought-movement in self-repellency; a going out each way in counter-negations; a being in what was known as the "universal negativity;" and in the development of such abstract thought-process, it assumed to determine all possible human cognition. It dealt only with facts, though transcendentally attained, and ignored all immutable principles. It concluded, from the facts found in us, what and how *we* must know, and was thus solely a critic of the *human* understanding, without attempting to determine any *other* order of knowing. We have in it a *critic* of human knowledge, but no *science* determining the validity of any form of knowing. It called itself rationalism, but is purely transcendental logic. It nowhere brings in the work of the comprehending reason, and uses solely the faculty of the connecting understanding. It is, at last, quite as empty as the scholastic logic, for it excludes all that is objectively real, and can thus never rise out of the sphere of the subjective ideal. That it should carry out and posit a valid objective, would demand that it should have the eternally real within its own subjective; but this is the Platonic, and not at all the Germanic transcendentalism. In making the understanding void at the beginning, no possible process of logical thinking can fill it at the close. It thus commences in a specious delusion, and terminates in a stupendous dream.

By surreptitiously raising the abstract and empty thought-process to a personality, and calling it the "world-spirit," the philosophy could elaborately disclose how this world-spirit educated itself to self-consciousness, and to know the universe as its own objective manifestation. The material universe and the spiritual humanity are developments of this absolute world-spirit, and the destiny and the immortality of man is, that he see himself identical with the absolute world-spirit, and that in the endless ongoing of the thought-process the universal mind is coming out, and the absolute is perfecting in self-consciousness, and in this is all the Humanity, and Deity, and Immortality, and Philosophy, that man can know. Surely Transcendentalism, though taking a longer road, and travelling through much more aërial regions, has hardly come out ahead of Common Sense, and done little more than any other school which admits only facts, to rescue Theology and Philosophy from Hume's skepticism or Comte's positivism.

Thus, ever, must the labor of the connecting understanding prove itself utterly incompetent for a valid theology or philosophy, and this as truly in the method of induction and of common sense as of the scholastic syllogism, and again as truly in a transcendental critic as in any other. It must think through a medium, and can never originate without something to come from, and something to put forth, and must thus have its Maker and Governor already made and conditioned. In the contemplation of its most profound abstractions, broad inductions, and transcendental developments, it is ever within the charmed circle of nature, and condemned to toil under the bondage of already determined conditions, and we are forced to cry—

Unless above himself he can
Erect himself, how vain a thing is man!

His progress is a cycle, and his path a tread-mill. A personal God is inconceivable within this sphere, and to this sphere there can be no conception of an outer and a beyond. Theology cannot begin, and philosophy cannot finish; for the first can find no Deity, and the last can find no link in which there is a reason for the whole chain.

We have, at last, the offer of ECLECTICISM, but it is not in a method to afford us any help. The name is here with no appropriate application. As the taking of what is deemed to be true from all other systems, Eclecticism must first have its own measure, or it cannot of right take any thing from any system; and when it has its own measure, it has already its own philosophic being and method. In other words, it already exists before any electing, and has its own law and method in order to any claim upon others. And in its author it has its well-expressed doctrine and method. Cousin's method dispenses at once with all transcendental analysis, and attains the absolute by direct consciousness. The human mind has the finite, the conditioned, the relative, immediately in consciousness, and to these, the infinite, the unconditioned, the absolute, are respectively correlatives. The first cannot be in consciousness without the latter, for indeed the first is nothing except in correlation with the latter. Just as the odd is nothing without the conception also of the even, so the finite, the relative, &c., are nothing without their conceived correlatives of the infinite, the absolute, &c. Where one is, the other must at the same time be. In the possession of these, we have also immediately their relation in consciousness, and can thus distinguish the one from the other. Given the *finite*, there is also at once with this given, the *infinite*, and the *relation* of the two; and in this apprehension of the infinite, the absolute, &c., we have the conscious knowledge of God.

All this is the spontaneous operation of the primitive consciousness, and thus belongs to all men in common, and may be known as reason. But in analyzing this operation and its results, each man goes about it in his own way, and each may have his peculiar opinion. Reason is thus common, impersonal, true; reflection is particular, personal, fallible. Taking the veracious reason, it spontaneously gives the absolute immediately and necessarily with every relative that comes into consciousness. Cousin takes causality only as the ground of his relative, and thus the relative cause at once gives the absolute cause, and this absolute cause is the Deity. He further proceeds, by saying, that causality is nothing except in action, and therefore the absolute cause must act, and go out into effect. The universe is as necessarily from the Deity, as the Deity is necessary to the universe. And, further, while God must go out into effect, yet does he not exhaust himself in the act of manifesting himself in objective effects; He is, and the universe also is, and such duality, he argues, excludes Pantheism.

But this reason, or spontaneous consciousness, is still only the faculty of a connecting judgment, and, as before said, can never attain to a legitimate theology or philosophy. The delusion is easily made transparent. Because the finite and relative *suggest* the Infinite and Absolute, it is thereby said that we *know* the Infinite and Absolute. A suggested conception becomes a cognition. But beyond this is the deeper delusion, that we attain a true Deity in this conceived Infinite or Absolute. Suppose we take the finite as applied to time or space, and let this be supposed truly to give Infinite time or space; is such Infinite the Deity? Or again, suppose we have the relative phenomena, and these suggest, or even validly give, the Absolute substance; is this Absolute substance the Deity? According to the philosophy, both Infinite time or space, and Absolute substance, should be God. But causality is actually taken, only because causality may the better be taken as Creator and Ruler, than either time, space, or substance. Take then Absolute cause, and is this the Deity? Not at all. It is only a conception of the logical understanding, and has its inherent conditioning just as truly as any second cause proceeding from it. It must go out into effect; yea, into just the effect determined in its own conditioning. It is cause caused, though arbitrarily termed absolute cause. There is here no personality; no capacity to originate; no self-determination; nothing of the supernatural. The absolute cause is nature still, and has in it its conditioned constitution, and we could never love and worship it, nor think a universe as coming from it, except as itself a part of it. Such a method can by no possibility reach to a true theology or philosophy.

The Platonic philosophy had the conception of God, as the Good, and thus as moral personality, and not at all as absolute substance or absolute cause. In this conception there was occasion given for the cognition of God as supernatural, while the restricting of the conception to substance or cause, though absurdly applying the term Absolute, necessarily confined it still within nature. God must be author of all substance and cause, and can himself be restricted by the conditions of no substances or causes. His conditionings can only be from the rational claims which spring eternally from his own rational being. What it behooves him to do as due to his own glory, or supreme excellency of being, that only can determine his action, and not at all the constituted nature of a substance or of a cause. Divine revelation has widely diffused the conception of a God, absolute, personal, supernatural; who originates the natures of all things "according to the counsel of his own will," or, which is the same thing, according to the claims of his own rationality, without himself being subjected to any nature. He looks only to the Archetypes essentially within his own rational Spirit, for the direction of all his creative and administrative energy. And it is a marvel and a reproach, that with all this Platonic and this Christian teaching, the world's philosophies are, to-day, all radically materialistic; holding all being as fact, or constitutionally natured; and are thus necessarily, in the end, Atheistic or Pantheistic. Seen from a comprehensive point of vision, they invariably and inevitably lead logically out to a complete exclusion of an absolute, personal, supernatural being from human knowledge and even from human conception. The reason of universal humanity calls for, and acknowledges, an unbegun, unmade, and supernatural Beginner, Maker, and Finisher of all that has a nature; and the Christian heart worships a Jehovah, whose sovereignty and authority lie underived and solely in the absolute behest of his own reason; while all speculative philosophy has come to ignore and deny every conception which cannot be brought within the connections of the logical understanding and subjected to the determinations of some constitutional nature. The conception of a Being who may begin from himself, and create objectively to himself, without finding himself caused to do so by any previous conditioning, seems utterly to have fallen out of all philosophical intelligence. Where is the philosophy, which can logically from its method, present a God to our acceptance as a *causa causans*, without being thoroughly a *causa causata?* Who seems to feel any shock at the absurdity and impiety of talking about the *nature* of God, and the *nature* of the divine will, as if the awful prerogatives of the supernatural could be brought and bound within the conditions of the natural? Our religious consciousness is clear and complete for an absolutely supernatural; our philosophic consciousness is, dogmatically or in its own supineness, trained to the restrictions of a relatively conditioned nature of things. It is among the strongest evidences of the deep and permanent working of the immortal reason within the soul, that notwithstanding the wide-spread prevalence of a philosophy everywhere sinking the Deity to a fact, there is yet the growing power of a religion which worships him as an unmade Spirit, in spirit and in truth. How much more rapidly may the knowledge and the worship of the true God spread, when philosophy herself shall become converted to, and baptized in, a Gospel theism!

What then we need for a truly *rational theology* is the conception and complete recognition of an absolutely supernatural Being—a God for the rational soul, and not conditioned to the physical necessities of the logical understanding. Such a demand met is sufficient for theology, and a valid answer to the perfectly logical Skepticism and Positivism before stated. Such theology may then be safely laid as the starting-point for a true *rational cosmology*, and in which may be embodied a thoroughly comprehensive and conclusive philosophy. In this way only is a valid theology or philosophy possible. In this way nature may be fairly presented as subjected to the determining conditions of immutable principle, and thus the facts of nature come to be known in their inherent laws, and having an eternal reason why thus they are, and not of some other nature. So matter itself may be expounded; so all the laws of motion, of gravity, of fluids, of falling bodies, of magnetism and electricity, of chemical and crystalizing agencies, of the ensphering and revolving of suns and planetary systems, and of the superinducing of vegetable and animal life, and of rational intelligence, may be interpreted. The full insight of all nature's facts, so as thoroughly to read all nature's laws, will not at first, nor very soon, be attained; but enough may be presented to give assurance that there is a rational philosophy of nature, as there is a valid theology above nature, and that we have started on the right path to find and finish it.

RATIONAL COSMOLOGY.

GENERAL METHOD.

THAT we may attain to a rational idea of Creation, it will be important that we first attain a rational idea of a Creator. Creation is an origination; something made where before there was nothing; and the universal Cosmos is inclusive of all that is so made. The Creator himself must then be without origin, and inhabiting eternity. The Cosmos is also a creation, beautiful and orderly, fashioned according to the determinations of immutable principle, and moves onward to a proposed and purposed consummation; the Creator must therefore be its Governor and Finisher, as well as its Author. He must originate all, and guide all that he originates to the end proposed.

Now the conception of such a Being is neither readily attained nor easily expressed. An absolute Author and Finisher, who encompasses all things before and after, while he himself is encompassed by nothing, is necessarily incomprehensible to a finite understanding, and can in no way be subjected to logical thought. No faculty can take cognizance of such a being but the insight of reason alone. The attainment of the idea involves all that has been termed "the Philosophy of the Infinite"—the Problem for finding the Absolute—and which by some of the most honored names has been denied to be at all accessible by the human mind. Others have deemed the problem to be of practicable solution, and have made labored attempts to accomplish it, while, under the delusion occasioned by the use of the wrong intellectual functions, they have only produced specious absurdities, or run out abstractions to utter negations.

A position can be attained, from whence these false methods of dealing with the problem may be seen in the necessities of the case to be futile; while by employing the appropriate faculty, the attainment of the idea of the Absolute may be completely successful. In a former work of Rational Psychology, a more extended examination of the subject has been made than is here needed, but inasmuch as a clear idea of an absolute Creator and Governor is conditional for all intelligent approach to a Rational Cosmology, a concise and independent mode for its attainment will be here presented. This will occupy the space given to the *First Chapter*.

Having thus the clear idea of the Creator, we shall be prepared to enter upon a detailed effort to attain a comprehensive idea of the creation itself. If the Creator must make and guide the universal cosmos after the determinations of immutable principles, so that his work may be truly fact pervaded by principle, then must the great plan have already been laid in the reason, as the archetypal idea of the whole making and finishing. To no finite reason, is it to be anticipated, that this plan will ever reveal itself in all the clearness and completeness of the divine Ideal; yet nothing hinders, since such a plan certainly is, that the human reason may not earnestly and reverently apply its powers to the attainment of its grand outlines, and in the teaching of eternal principles find, by a rational insight, what and how creation must have been, and read her great laws, not as mere arbitrary facts, but as the necessary result of a work rationally begun and wisely accomplished. This will fill, at much greater length, the *Second Chapter*.

When the Cosmos is attained in its plan and principle, it will be necessary to take the facts as actually given in experience, and study them with the direct design to find their law as plainly determined in the eternal principle. Facts teach nothing until they are seen in their principles; but when the principle is applied to the fact, and the fact is read and expounded in the principle, then have we, and only then, a rational philosophy. This will be the work for the *Third Chapter*, and which might be prolonged indefinitely.

CHAPTER I.

THE IDEA OF AN ABSOLUTE CREATOR AND GOVERNOR.

The human intellect has three different functions for knowing, in each of which the processes pursued and the cognitions attained are different in kind one from the other, and no supposable augmentation of degrees can bring them to become identical. An imperfect analysis, which fails in the psychological recognition of these three different kinds of knowing, among other imperfections and errors, will inevitably exclude from all intelligent approach to the question of the Absolute, and oblige to the denial that any such conception can be legitimately sought by a finite mind. These three distinct functions of intellectual agency are the Sense, the Understanding, and the Reason. An Absolute may be sought in them all; the true Absolute can be conceived and attained only in one; the nature of the case *à priori* determining that, to both the functions of the sense and the discursive understanding, all attempts towards the conception of an Absolute involve an absurdity, and must therefore ever rest under an utter impossibility, while the reason is directly competent to state and expound the whole problem.

The intellectual agency in the sense performs its work and attains its end only through a process of *conjoining* the manifold into unity, and thus constructing the indefinite within limits. This agency in the understanding works to its end only by *connecting* the separate and fleeting into a permanent, and by this discursive process concluding in judgments. In the reason, the intellectual agency attains its end by an immediate *insight*, which detects the necessary principle that comprehends the universal within it, and in this compass of all that has limit and relation at once attains and recognizes the Absolute. If we make these processes of intellectual agency cursorily to pass beneath our inspection, we may clearly determine in the cases themselves why the first two cannot reach to an Absolute, and how the last can both attain and expound it.

1. The Absolute as the Infinite.—The mere sensation in any organ can be only a content given for a perception, but cannot complete the perception in any case. An intellectual action is necessary first to distinguish the peculiar sensation and thereby attain the quality, and then to bring the whole within limits and thereby determine the quantity. Quantity may have limits under three general modifications, viz., limit in space, and thus shape be perceived; limit in time, and thus period be attained; and limit in the intensity of the sensation, and thus the amount of the quality be known. No matter how distinct the quality, the quantity must also be made definite or we cannot have a clear and complete perception. We may observe the colors distinctly on the page of a book, but without an accurate defining in shape we shall not know the etters.

This intellectual act of defining may be made a purely subjective operation within the mind alone, and the limits of degree of intensity would then be excluded, as having relevancy only to some actual sensation, leaving only limits which determine the definite shape in pure space and the definite period in pure time. But such constructions within pure space or time can be possible only in one method. The intellectual agency must go through the contiguous points in space and conjoin them into a line, and thus carry the line about an area, or it cannot define any pure shape; and must also go through the consecutive instants in time, and thus begin and terminate a duration, or it cannot define any pure period. In other words, the intellect can possess no definite forms in either space or time except as it constructs them itself by its own act. Pure space and time will not have any limits in them, but only as the intellectual agency makes them.

Now the Absolute, in either space or time, must be a whole which cannot be carried out any further, and is thus absolved from any further modification. It is either a whole *so small* as not to be capable of further diminution, or a whole *so large* as not to admit of further augmentation. From the very necessity of the case, the conjoining agency that constructs within limits, and thus determines a completed whole, must itself describe the boundaries and carry its own lines entirely around every form that it attains. It can have no figures that it does not itself describe, and no periods that it does not itself limit by both beginning and ending. We will then put such a conjoining agency upon its search for the Absolute in the direction for finding a whole *so small* that it cannot be further diminished. The insight of the reason may enable the intellect to say of a circle, for an example, that there must be a point in it which has no radii, but in which all the radii of the circle terminate; or, that if that circle with its area revolve, there must be a point in it which does not revolve and which can thus have no upper nor lower portion of the circumference; and in each case there must be involved the conclusion that here is that which cannot be further diminished. As a supposed absolute, the intellectual agency may set itself in this direction to construct and thus to possess an absolutely least whole. In order to its attainment as a whole it must construct it within limits, and must either begin within and go out to and around its circumference, or begin without and describe its circumference around its centre within, and can never possess any completed whole, however small, without thus drawing lines about it. But no such whole can be the absolutely least, for it must have its upper and lower portions of the circumference, and be capable of revolution, and possess radii. The Absolute, to the insight of reason, in this direction, is not thus of any constructed whole as a limited, but only of a limit; a point between upper and lower portions of a circumference, or a point neither out of nor within the radii, but the limit where they terminate. The Absolute, thus, cannot be conjoined as a whole for the sense, but is necessarily to it THE INFINITE; that at which the conjoining agency may begin, or that at which it may finish, but that at which it cannot both begin and finish. The intuitive Absolute must thus be indefinable by the only function which the sense can employ; and the absolutely least, as a whole so small that it may not become smaller, is to the sense an utterly unattainable cognition. Its smallest must be wholly constructed within limits, and ever that which is limited the limit can divide. The same is true of period as of place, and thus the attainment of a whole as absolutely the smallest is necessarily impracticable.

We will again put this conjoining agency upon its search for the Absolute in the direction of attaining a whole *so large* that it cannot be augmented. The comprehending reason may say of space, that there must be a whole of immensity which is not any part of itself; and also may say of time, that there must be a whole of eternity which cannot be any of its parts; and the intellectual agency may go forth to construct this absolutely greatest whole. As before in diminution, so here in augmentation, the conjoining intellect can have no whole which it has not completely surrounded by the line that itself carries, and thus can know no whole of space without completely limiting all space, and can know no whole of time without completely bounding eternity. But the greatest definite place is not yet all space, and the greatest definite period is not yet all time, and thus at the furthest augmentation there is more beyond, and however much may have been defined, yet of both immensity and eternity still we must say, that each is THE INFINITE; that which it is impracticable to finish.

As now, the only fields in which a conjoining intellectual agency can work are those of space and time, and as in neither can an absolutely least nor an absolutely largest be attained, it is quite manifest, from the nature of the case, that in no way to the sense can the Absolute become known. The forecasting reason postulates both the absolutely least and largest, but when the constructing sense sets out to execute the work, it ever finds itself with the infinite beyond, and never that it is at the Absolute. There is an intrinsic antinomy in the human mind; a law that demands an Absolute, and a law that forbids it should be found; and till an accurate analysis has discriminated and thus reconciled the different functions of knowing, the mind is really a riddle or an apparent absurdity to itself. To the sense there can never be an absolute whole, either the least or the largest; there still ever remains to it only the Infinite.

2. THE ABSOLUTE AS THE UNCONDITIONED.—When we have put the quality completely within limits, and thereby made it to stand out in consciousness as a definite whole, the intellectual function in conjoining or constructing has done all its work, and the product is a perception, or a phenomenon taken through sense. The definite qualities are thus known, but these qualities are so known only as separate and fleeting appearances. The sense can affirm what thus appears, but cannot at all think the appearing separate qualities to be the attributes of some common subject. The sense can give definitely all the qualities which belong to the rose, but in the sense they are separate qualities only, and no subject, as the rose, appears in the sense at all, nor any act which puts the qualities together into the subject rose is at all put forth by any intellectual agency in the sense. There must here be introduced altogether another kind of agency than that which conjoins within limits. We need for this the *connecting* agency of the understanding, and which is explained in the following manner:

The sense can take no cognizance of a *substance*, but only of *the qualities*. The qualities appear, the substance does not appear. I *think* that some existing thing has impressed the organ of sense, and thereby has given a sensation which I have discriminated and defined, and in my *thought*, I refer this distinct and definite appearance to that thing as its subject; and then in as many ways and through as many organs as that thing is thought to give distinct and definite phenomena, I successively conclude these phenomena to belong to it, and thus judge discursively the qualities to be predicated of one common subject; or, which is the same thing, the qualities of one common substance. The sense gave the qualities distinctly and definitely, and then quite another intellectual function intervenes, and, taking each quality discursively through the same substance as given in thought, connects them all in it by judging them all to inhere there together. The qualities are thus no longer separate, but the attributes of that one substance, and these qualities thus connected in that one substance, are known henceforth as one thing. By thus thinking in judgments we come to know that the sense phenomena have their common ground in the one substance we have thought for them, and the intellectual function, by which we have been enabled to connect these qualities into one thing by making this substance to *stand under* them, we term *the understanding*. From the nature of the case, it is thus impossible that an understanding should work in connecting qualities into things, except as the notion of substance is given to it; the moment the thought of the substance is lost, the very medium of all possible connection of the qualities would be gone.

Still further, the qualities as given in the sense often vary in the same ground, and the one thing changes its appearance. The hardness and brittleness, &c., of the ice give place to the fluidity and limpidness, &c., of the water. We think these last qualities as still in the same substance, and thus know both the ice and the water to be yet one thing, and yet we think that one substance to have been so modified by the presence of some other substance, that the old qualities were made to pass away, and other qualities as new events to come out from the same source. The passing away of one and the coming of another event is given in the sense; the sequences appear; but the modifying efficiency of the one substance upon the other does not appear. This is *thought* only, and the discursive process again brings the successive events into connection through these modifying combinations of substances, and knows the modifying efficiency as cause and the modified event as effect, and thus judges the sequences to be a linked and orderly series. The understanding, again, must have this notion of cause, or from the necessity of the case the only medium for connecting the sequences into a linked series would be lost. All qualities are thus judged as inhering in some substance, and all events as adhering to some cause, and thus the separate qualities of the sense and their changes are bound in connection as one common nature of things, and all constitute but one world or universe.

It is now manifest, that in this field of the understanding, as before in the field of the sense, the occasion is given for seeking after the Absolute, though in quite a different form. Not the Absolute in reference to any limited and completed whole, whether least or largest, but the Absolute as the substance which has nothing deeper, or the cause which has nothing higher. We thus put this connecting agency upon the search for the Absolute, in the first place, in the direction of the absolutely deepest substance.

The reason may intuitively say that some substance must be the ultimate ground on which all substances rest, and the discursive understanding may be put upon the start actually to attain to it. These qualities have been judged to inhere in a common substance, but on what does this substance rest? It cannot be self-supported, for from the necessity of the case a connecting understanding must have the medium through which the discursive connections are to run, and so soon as you leave the substance to itself it hangs as helplessly over a void as would a quality without a substance. The understanding must, therefore, think this substance as some modification of a deeper substance, and if it would reach the Absolute by thinking in discursive judgments, it dooms itself to an endless descent where each dropping footstep can only fall upon a stair that must be conditioned upon another yet beneath it. To attempt the conception of a substance originated, or of a substance annihilated, is the absurdity of connecting without a medium; of thinking a substance that was itself unsubstantial, or of thinking away a substance that yet should leave all above it to be substantial. It would cut off the thought from all possibility of connection, and the discursive understanding can look at this only in horror and helplessness. An absolute substance is thus manifestly unattainable, and could be conceived only as an arbitrary stopping upon some one as an ultimate, but which yet, by the very necessity of the thinking function that demanded this for all above, demands for this yet also another beneath. An absolute substance would be THE UNCONDITIONED; the substance that stood under all others with no substance under it; but such conception of the unconditioned could not also be a conception of the substantial. The absolute substance is necessarily to the understanding an absurdity; a contradiction to the necessity of thought; and can therefore never become a cognition to the discursive intellect. The true Absolute is as remote here from an unconditioned substance, as before in the sense from the infinite in space.

In another direction, the discursive intellect may be put upon the search for the Absolute cause. The reason may affirm that there must be a cause which is the source of all causes, and as thus itself uncaused is an absolute cause, and the understanding sent on the way after it. But, again, from the necessity of a discursive process, the medium of connection must be maintained, and the attempt to stand on any cause arbitrarily assumed to be the ultimate in the regressus, or the first in the outgoing of the following series, is the putting yourself with one foot on a retreating stair while the other vainly seeks to plant itself upon vacancy. The Absolute cause is for the understanding an unconditioned cause; a source of all causes with no condition above itself what shall come out of itself; and is thus the absurdity of a source for all efficiency with nothing to make itself effective. The true Absolute is as diverse from an unconditioned cause here in the understanding, as it was before in the sense from the Infinite in time.

There is here as manifest an antinomy in the human intellect as before in the sense. The reason forecasts and postulates a substance that has no substance beneath it, and a cause that has no cause above it, and yet the very function of the discursive understanding forbids that such an Absolute should be cognized or even conceived. There is an insoluble paradox from the very working of the human intellect, except as in our psychological analysis we have found that the function demanding and the function forbidding are entirely distinct in kind, and that each is to be held responsible only for its own cognitions in its own processes.

3. The Absolute as in the Understanding itself.—There is also another method of attaining to an Absolute, which takes the understanding itself, and transcending the consciousness, in which is all our ordinary experience, carries out an analysis of the understanding, as the function of judgment, to its constituent elements, and finds an Absolute in the understanding itself. This is still a use of the discursive faculty, and only turning its action upon the constitutive elements of its own being instead, as before, upon either space and time in the sense, or upon the notions of substance and cause in the thought, and instead of an absolute whole so small as not to be diminished or so large as not to be increased, or an absolute substance or absolute cause, it assumes to find an Absolute in the understanding itself.

Beginning in Kant and passing through the speculations of Fichte and Schelling to Hegel, we have the following modifications of this method of finding the Absolute in the understanding itself. The human understanding is taken as the faculty for thinking in judgments, and is originally constituted to possess certain primitive conceptions which become the general forms for all varieties of logical judgments, and which are thus termed the categories of the pure understanding. These primitive forms, with which our human understanding is constitutionally endowed, determine and limit our whole sphere of knowing, and when analytically formed they enable the transcendental philosopher to say beforehand, from the very constitution of the faculty of judging, what is the entire capacity of man for attaining cognitions. He can know in all the forms provided for him in these primitive conceptions, and can conclude in no judgments which do not range themselves under some one of these categories.

Above the general forms for concluding in judgments through logical syllogisms, there is also a constitutional provision for directing the ascent from the major premiss of one syllogism to the conclusion of another on which it has depended. The major premiss of any logical syllogism must be an assumption, except as it has been deduced in the conclusion of a pro-syllogism; and to prompt and direct the mind along this ascending way up the ladder of syllogisms, there is the higher primitive conception of the Infinite, or the Unconditioned, constitutionally given to man, and which, as the subjective Idea of the Absolute, regulates this logical regressus in the same manner as if a real ultimate might at length be reached, beyond which there would be no occasion for a pro-syllogism. As the primitive conceptions for single syllogisms were termed categories of the pure understanding, so the primitive conceptions in the various processes of rising to the Infinite, the Unconditioned, and the Absolute, were termed the categories of the pure reason.

The understanding and the reason are thus only different varieties of the same logical function for concluding in judgments, one regulating its process by given primitive conceptions in the simple syllogism, and the other regulating its process by a higher grade of primitive conceptions in the ascending march through indefinite pro-syllogisms. The reason is still discursive, and not at all the immediate insight of the Platonic Reason. The Absolute is here a primitive conception; a regulative form of thought in the subjective understanding; and thus an ideal Absolute only. Whether there be a veritable ultimate or not can never be determined by the human mind, for it can only regulate its search for it by this subjective ideal Absolute, and can never reach it. The true Absolute is wholly problematical; it can neither be proved nor disproved; the ideal Absolute constitutionally given to the human mind is all that can be cognized by man. We are so made that we think an Absolute, and thus regulate our ascent toward it; but we can never attain to it and plant our logical footsteps upon it.

This Kantean Idea of the Absolute *in* the Understanding became subsequently transposed for an Absolute Understanding itself. That agency, which works out in consciousness the ego or the self that we know, must be back of the ego or self which is known, and cannot itself be brought up into the light of consciousness. The self which we come to know is the intellectual product of a deeper self which we cannot make to appear. This deeper self works up into consciousness a self which is then known as subject, and also that which is distinct from self, a not-self, which is then known as object, and thus our whole experience of subject and object has a deeper source absolutely independent of all conscious experience. This underlying self which develops itself into the conscious self, and also into the conscious not-self, *i. e.*, into both the subjective and the objective, is also altogether the source for each personal self in its separate consciousness, and therefore out of it come, and in it are identified, all separate self-consciousnesses, all subjective, and all objective experiences. One absolute self is the germ that evolves itself into distinct personalities, into conscious subjective experience, and into the conscious experience of all that is objective; and hence the whole intellectual life of humanity is but an outgrowth from an Absolute, which is back of, and beyond all possibility of appearing in, consciousness, and which can be known in no way but by "an Intellectual Intuition," which penetrates beneath the subjective consciousness and beholds it face to face. This Absolute Ego is taken to be a real, acting, self-evolving being; the identification in himself of all that comes to have existence; and all existence is, in fact, only the stating or positing of his perpetual self-evolution. An Absolute real understanding thinks out into subjective and objective existence all that is known.

Ultimately, this Absolute real being, thinking itself out into personality, subjective consciousness, and objective experience, becomes thoroughly dissolved into an utter abstraction, and there is no longer a self, or ego, as a veritable understanding, but solely a *thinking process;* not any substrate agent, but merely a living movement; and this pure thinking movement is assumed as the Absolute, and by a law of perpetual dialectics, or reciprocal counter-negations, works out the universe of unconscious matter and self-conscious mind. The Absolute thought-movement, beginning in abstract being, which as entirely abstract has no distinction, and is thus identical with naught first denies or negates that it is naught, and then by a counter-negation denies that naught is being, and thus posits being as no longer abstract being, but as being excluded by naught; and thus in this diremptive, or counter-negative movement, abstract being has come to *stand out* with naught over against it, and each mutually excluding the other, and is thus the thought of *existence.* But the thought-movement cannot rest in existence; it goes out from existence and negates any limit, and thus thinks the Infinite, and returning to existence it again negates the Infinite as limited in existence, and thus thinks the Finite; and in this counter-negation of the Infinite and Finite, existence has become in the thought not merely a being as standing out from nothing, but being as every way limiting itself, and is thus being *per se.* Thus the living movement is traced through a perpetual series of counter-negations, each one conveying the thought further on and positing a new cognition, till the thought-process has given in its course all of nature, educated itself to self-consciousness, to universal intelligence, and at length to divine Omniscience. Thus Kant's Absolute is a subjective regulative thought; Schelling's Absolute is the infinite understanding in its original germ; and Hegel's Absolute is an abstract thought-movement which has not yet posited any thought, but in its endless ongoing is at length to state existence, nature, personality, humanity and developed Deity, and come at length to know itself as the subject of whatever is, and the object of whatever itself knows.

Of this whole transcendental method for finding an Absolute, we can say, from the necessity of the case, it must be unsuccessful. It uses only a discursive faculty and employs only the processes of analysis and abstraction. It begins in experience, and analyzes and abstracts till it assumes to find that in which experience is conditioned. But this root and source for all experience is still a constituted being; a something given with its own necessitated law of action imposed upon it; and even when the abstraction has gone beyond all substrate being, and retained only a movement in which there is nothing moving, it still must come under an intestine necessity, and work according to a constituted nature, and subject itself to conditions it already finds within it, and above which it can never exalt itself, and from which it can never deliver itself. It transcends experience, not by going back to eternal principle which must determine all experience, but simply by going back of human consciousness and finding the constitutional elements which regulate human experience in consciousness, and thus determining what our human understanding can know, simply because it finds that we are made with functions that primitively capacitate us to know thus, and not otherwise.

Not that which is above fact and nature, but that only which is above human consciousness is sought, that in the end it may attain a constituted *principium*, a created source for all that conscious experience has given. This the discursive understanding can very well accomplish, for it is only undoing its own work and raveling out the thread that it has knit. Having put together by a connective process in judging, it may readily unweave its own web, and go back in abstraction towards nihility, until nothing be left but the mere semblance of any content in the thought, and then by terming this highest abstract element the Absolute, it may readily begin with it and retrace its old process of putting things together, and in this assume that the ideal work is creation, and the empty product a universe. But from the nature of the process, if it stop short of utter annihilation, the highest abstraction must be still something that the intellect had when it began the analysis, and can be no more an Absolute above and beyond nature than was the whole furniture of its thought when it began the abstraction. By analyzing and abstracting from the conditioned we are making no progress toward an unconditioned, and an endless analysis and abstraction of the understanding can never find an Absolute.

We may readily cheat ourselves by calling this abstract thought-process the world-spirit, as if it possessed a valid being, and then call the empty thinking the development of this world-spirit, and delude ourselves as if we had built over in our thought that which the world-spirit had actually posited and stated in its ongoing ; and yet even this delusive creator and creation would be a thoroughly finite and conditioned conception. This assumed world-spirit can only act in one way, and go out in one process, and take one step in its perpetual counter-negations at a time, and all this with no final end to be reached, and no free purpose to be attained, and no approving inward consciousness to cheer and reward it. All is thoroughly within the conditioned understanding, and is but nature still, and grows up under as rigid a necessity as the tides flow or the planets roll. It seeks to transcend sustained substance and conditioned cause, but this abstract Absolute is still grounded in the substantial and bound within the causal, and utterly helpless without a something on which to stand, and a supplied efficiency from whence to draw.

All philosophy of nature is hopeless and helpless without the full recognition of the absolutely supernatural, for all exposition of nature in either its origin or its end, must be found in that only which is above nature. But every attempt to reach the supernatural and cognize the Absolute by any work of the discursive understanding, is vain. We may employ it upon the pure sense in the conjoining of space and time, but all possible constructions here will still leave the Infinite unconstructed, and can therefore never find an Absolute. We may employ it upon its own notions of substance and cause, but all possible attempts to descend to an unsustained substance, or ascend to an unsupplied cause, can never stop in any one substance or cause which is not conditioned already in its own being, and thus leaving the unconditioned wholly beyond its furthest march, and of course the Absolute yet unattained. Or, we may lastly set the understanding to work upon its own functions, and analyze itself up to the primitive elements which enter into its original constitution, and attain its most sublimated transcendental abstractions; but we can never take that in the end which was not also given to us at the beginning, and from the very fact that it was originally comprehended within the understanding, it must be impossible that it should ever become the compass for comprehending the understanding itself. It must ever be the included and can never become the absolutely conclusive. If, then, we have not the endowment of some distinct and superior function of knowing than the discursive understanding, we are from the nature of the case shut out from all entrance upon the field where lie the problems of the Absolute. We are doomed to wander up and down through the connections of nature, and can neither know nor conceive any thing of the supernatural. It is certainly very much gained in the saving of severe but fruitless labor, to know that no conjoining and no connecting intellectual agency can be at all used in the philosophy of the Absolute. It is more gained, to know that we do not need any such aid. Neither Absolute time nor space, neither Absolute substance nor cause, neither a transcendental regulative Ideal Absolute, nor an Absolute thought-process, could bring us to the being we want.

4. THE ABSOLUTE AS GIVEN IN THE REASON.—We cease, then, altogether from the use of the discursive understanding in this work of attaining the Idea of the Absolute, and attempt nothing through completed constructions in space and time, nor connected judgments in substances and causes, nor analyses and abstractions of the function of judgment itself. We have a position from which we see that all such labor must be fruitless, and we turn to the use of the reason solely; the faculty for direct and immediate insight. That we have such a faculty, distinctive in kind, and giving to us all our prerogatives of rationality, personality, and free and responsible originality, is sufficiently clear in the consciousness of its own working. In pure diagrams we see universal truths without any process of logical deductions, as that any three points in space must be in one and the same plane; and that any two sides of a triangle must together be greater than a third side. In pure physics, we see that action and reaction must be opposite and equal; and that compound forces must give their conjunct direction to motion. In pure forms we can see spiritual sentiment, and thus have an ultimate standard

of taste in the beautiful; and in spirit itself we can see an intrinsic excellency that demands for itself that it should be end, and not means to an end, and thus have an ultimate standard of right in the good. We will apply this rational insight to a series of grounds, in which may be detected the working of other than material forces; and also to the distinctions in an ascending spiritual spontaneity up to the supernatural; and in the supernatural we will detect also the point which separates the conditioned from the unconditioned, and come directly upon the Absolute and Divine.

Let it be here remarked that the Absolute we seek is not excluded from all relations and conditions. That which should be utterly without relations could not be expressed, and that which should be utterly without conditions could not be explained. It is only necessary that the relations and conditions should be wholly subjective, self-directed, and self-sustained, and bringing with them no dependence upon nor amenability to any outer being. Not without self-relations and self-conditions, but wholly absolved from all dependent relationship and subjected conditioning to any other. We proceed then on our course to the complete attainment of such Idea of the Absolute.

A grain of wheat may be wrapped up in the same cerements together with an Egyptian mummy. Thousands of years pass away, and not a moment in the long period has been without action neither in the living wheat nor the dead mummy. But to the insight of reason, a broad distinction is seen between these perpetuated activities. In the dead all the agency has been from without, and coming upon the subject that has been modified and changed by it;

while in the living the agency has been its own, springing up ever fresh within it, and resisting the outer agencies that would corrupt and dissolve it. The one has been the mechanical attrition of material forces, the other has been the spontaneous spring of a living energy. We seize upon this vital energizing, as reason gives it to us, and reserve it for our purpose in our future progress.

This living energy can only act according to conditions imposed upon it. It cannot germinate and propagate itself in new grains without the air, the earth, the sunshine, and the moisture. It is a power put into matter, and which has the capability to control and use matter, but only according to conditions imposed upon it, and when these conditions are supplied it is still conditioned within itself, and must grow out after its own controlling law, "first the blade, then the ear, afterwards the full corn in the ear," and this full corn in the ear only the "seed after its kind." With all the spontaneity of life, the vegetable is still bound in matter, and even its life is conditioned by an imposed law which it can by no means transcend; and thus its whole being is in and of nature only.

The ox that treads out and eats the grain has all this living energy, with the very remarkable addition that it can feel itself and give back sensation for sensation. Through the power of sensation it can be impelled to locomotion, direct itself in the selection of its food, and guide its experience by rules of prudence. The insight of reason finds at once in this a higher grade of spontaneous energizing, and knows that here is an approach towards self-direction. The animal can condition itself by its own sensations. We take then this higher idea of spontaneity which reason has gained in animal life and hold it for our purpose.

But this animal life and sensation is also in matter, and subjected to all the conditions of matter. Its very sentient life, which distinguishes it from the vegetable, is active only through matter and towards matter. It uses and seeks the material only, so that if we speak here at all of spiritual being, it is of "the spirit of a beast which goeth downward." Its feelings are all determined for it in the laws imposed upon it; and the sentient life can neither assume nor propagate other laws of energizing than those of its own kind. The animal is therefore yet wholly in nature.

Man has, beside the sentient animal life, the far higher endowment of a rational existence. The peculiarities of his rational being are in the following distinctive elements. He can originate for himself what to him are the perfect ideal patterns or archetypes of that which is the beautiful, the true, and the good, and use these to measure, criticize, and estimate all that experience may offer. Not what is taken from experience, but what his own genius creates for him, is his criterion for testing what he shall approve and what disapprove. He has his own principles or standards of judgment within himself, and with which the material and sentient world has nothing to do. He has also that self-knowledge which determines the intrinsic excellency of this his rational being, and what is due to himself and worthy of himself in all his actions. He can thus feel the claims of self-respect and responsibility to his own conscience, and know the retributions of self-approbation or self-reproach according as his deeds sustain or violate the law which his own rational being imposes upon himself. Here are peculiar self-relations and self-conditions, all subsisting within the rational, and having no dependence upon his animal being. The rational activity is competent to guide and determine itself alone, both without and even against the animal life.

The beautiful, the true, and the right are in the reason itself, and instead of copying them from nature and experience, it judges both nature and experience by them. It can move itself not only without the promptings of sentient nature, but directly against and over them. All of nature may be on one side, and yet the rational can say, I ought and I will stand and act on the other side. It can make its own conscious worth and dignity its end of action, and exclude all other ends which nature may present from holding any competition with this. Here is a real spontaneity, related in its activity and its law, its going forth and the end it is to reach, only to itself. It furnishes its own end and occasion for its activity. Its references of agency are all within its own sphere, and its conditions of direction and result are all self-imposed. It is self-activity self-directed. It is not from nature nor subject to nature; it is wholly above nature. Allied to nature as it is, even as the human is the combination of the animal and the rational in one, still the rational is not lost and absorbed in the animal, but ever asserts its prerogative over it.

Take, then, this free personality; this spontaneous agency with its law written upon and rising out of its own being; and we have made a long advance in our way to the Idea of the Absolute. We have found that which may absolve itself from all the domination of nature and stand forth wholly supernatural. It is no product of the discursive faculty, and no attainment of analysis and abstraction, but a cognition attained only by the direct insight of the reason. The eye of reason sees in the ground of the human, that this self-activity and self-law is the very prerogative and crown of its being, making it competent to rule over nature, and to live immortal with no help from nature.

But truly an activity that goes out of its own accord, as is the rational in humanity, and thoroughly supernatural as it is, yet is it ever subject to the colliding influences of flesh and sense. Even when regnant over every appetite, its virtue is the result of perpetual watchfulness and struggle, and at no moment can it rest in serenity and be sure that some suppressed lust shall not suddenly spring up with inordinate clamors for gratification. It can never be otherwise than militant even if triumphant. It is in an enemy's country, and owes all its security to its sleepless valor. In fallen humanity the rational has already sold itself in bondage to the animal, and basely subjected itself to nature, and thus conditioned itself under a load of necessities that it is morally helpless to throw off. "The law in the members" continually "wars against the law of the mind, and brings in subjection to the law of sin and death," and thus with all its liberty and responsibility and supernatural activity, humanity is no ground in which to look for the Absolute. Individually and collectively, the race is still so bound in the conditions of nature, either by constitutional or moral alliance, that it is utterly vain to hope that it shall ever be thoroughly absolved therefrom.

There is an order of beings above us, brought to our knowledge by revelation rather than by any human experience, who in all the prerogatives of rationality, as self-active, self-directed, and self-rewarded, are endowed like ourselves. They find an ultimate end to their activity in the securing of their own worthiness of character, and a constant stimulus and directory in their own conscious self-responsibility. They know that they cannot stand before the tribunal of another, unless they can pass the scrutinizing ordeal of their own consciences. They bring up from within higher and wider and more perfect ideals of beauty and truth, and can thus criticize and estimate whatever may come into their larger experience, much more accurately and comprehensively than can be effected by any of the human family. They thus are higher in supernatural endowment than ourselves. Add to all this the great distinction that they are not incarnate, and have no subjection to the clogs and collisions of matter. Their reason is incorporeal; or if there be a corporeity, it is what the inspired apostle calls "a spiritual body," "bodies celestial;" and thus imposing upon the activity and the spiritual sensibility none of the chafing, fretting, tempting excitements of our carnal nature. Nor, in the case of the great mass of the unfallen, have they ever subjected themselves to the corrosion and desperation which necessarily accompany the remorse of conscious guilt. Here then is an absolution even from an alliance with matter, and also the much more important consideration that they are absolved from all the conditions, without and within, which sin imposes. Here is a spontaneity more elevated, and a law of higher import and in clearer characters written on the heart, than any that the eye of reason has yet before disclosed to us. Surely in so wide an absolution from all outer conditions we must be near the attainment of the Absolute we are seeking. We fix this conception of angelic completeness and perfection in our mind, and ask, is not here the absolutely free, good, and holy?

But when the eye of reason looks more narrowly and penetrates further within the sphere where angel and archangel dwell, many occasions of outward restraint and conditions of imposed necessities appear that divide the angelic world by a broad line of demarcation from the sphere of the Absolute. There are ranks and orders among them, giving rise to the distinctive appellations of archangel, thrones, dominions, powers and principalities; there are perhaps different kinds of being indicated by the names Angels, Cherubim, Seraphim, and the Living Ones; all imposing the conditions of superior and inferior, prerogative and subordination, the tie of a class and the spirit of a party; and thus opening the door to the incoming of selfish spiritual passions in all the shapes of pride, envy, hate, jealousy, ambition, oppression, rebellion, deception, &c., and which have actually entered in all the malignity and mendacity of the fallen spirits. The highest angel is still limited in rational powers and activities, and thus necessitated to often come upon themes of speculation, and questions of practical interest, where he cannot see what is truth, or know which is duty, and thus in the necessities of his finiteness he finds himself conditioned to seek light and receive commands from one that is higher than himself. He cannot absolve himself from these conditions, and though he were to be true to all the light he has, he cannot rise above and free himself from these necessities, but would be doomed to go on eternally in darkness, doubt, and fear, from which nothing that he could bring out from himself could deliver him. In his very elevation, the boundless unknown yet before him would leave him as really helpless as the lowest, while the magnitude of the issues to his errors would make him more fearfully dependent than any. The highest angel must be ruled by a higher, and thus conditioned to an outer authority, or, in his finiteness and ignorance without a higher, must be doomed to the conditions of more terrible necessities, and thus in all ways must he stand far below the Idea of the Absolute.

Now, in all self-activity and self-law there is the conception of *spirit* as opposed to matter; essence which is not substance; and also is there in this involved *personality* as opposed to mere animal identity; a law both in precept and penalty *sounding through* the whole being as self-enacted and self-promulgated. The being is bound to be himself his own end, and cannot get his own or another's approbation if he permit himself to be made a means to any other's end. In humanity is such personality, but not pure. The man is not only personal, but also animal as thing in nature. The angel is pure personality, but not Absolute. He must depend on the instructions and requisitions of a higher. The authority of another may send the sound of a higher law through the soul, and then his self-law will also sound through his spiritual being, that in no way can he be worthy of his own approbation but in unquestioning obedience to this higher law. To that he is conditioned by what is due to himself.

What the insight of reason sees here to be necessary is a Supreme personality, elevated above all possible authority which can come from without his own being. The inner light of reason must exclude all need of instruction, and the intrinsic excellency must exclude all higher law. The Supreme Spirit is, thus, elevated above all outer authority, and absolved from all obligation *ab extra.* He is conditioned solely by what he knows in himself is due to himself. The Supreme Spirit is therefore Absolute self-law and self-determiner. This is the Idea of the Absolute in the reason; not at all the Infinite in space and time; nor the unconditioned in substance and cause; nor the purely abstract movement of thought itself; but a Supreme Spirit self-determined. Not without relation, for then he could not be expressed; but a pure spontaneity viewed in relation to its own known intrinsic excellency. Not without conditions, for he could not then come within any explanation; but conditioned only upon the perpetual behest of his own dignity, or, in other words, that he do all things for his own glory.

Such supreme self-determination is the very conception of Absolute Reason. All that belongs to nature is excluded from it. There is nothing to be constructed within limits, for it is independent of space and time; there is nothing to be connected as of qualities in a substance, or of events in a cause, for there is nothing to which the conceptions of statics and dynamics, physical substances and causes can have any relevancy. It is utterly supernatural, and nothing of the laws and conditions of nature can possess any significancy in reference to it. Reason is not a *fact;* a somewhat that has been *made;* but from its own necessity of being can be conceived no otherwise than a verity which fills immensity and eternity. In it is self-knowledge, self-action, self-direction, since it contains the archetypes or patterns of all possibilities, and the reason for taking which in the determination of such only as may be worthy of its own acceptance; and in this perpetual equalling and filling its own demands, there is necessarily perpetual self-satisfaction and self-approbation. It is unreasonable that there is no such end, and no such determined activity towards it, and certain attainment of it; or, which is the same thing, it is absurd to suppose that the Absolute Reason should not both *be* and *fulfil* its high behests. The conception of the non-being of the Absolute Reason involves the absurdity of conceiving reason to be unreasonable. To the Absolute Reason there must be the known necessity for itself; the conscious absurdity that it should not be.

Such Absolute Reason is manifestly a Person, having in himself the knowledge of all possible, and the self-determining will to execute all his own behests. To him there can be no beginning nor end, for there can be no time when he was not; and to him there can be no bounds, for there can be no place where he is not. He is unsustained and uncaused, for there can be no substance which he does not hold, and no cause which he does not originate. He is absolved from all dependence upon and determination by any being other than himself. Here is no abstraction, but the positive affirmation of the I AM; he who has being and blessedness and exhaustless fulness in himself; even the being of whom it would be an everlasting absurdity to suppose that he was not, and was not blessed, and was not satisfied. Sense cannot *per*ceive Him; discursive thought cannot *con*ceive Him; only a spiritual discernment, the direct insight of reason, can behold Him.

All the attributes which our manner of conceiving apply

to him participate in this characteristic of absoluteness. His wisdom is absolved from all dependence upon outward conditions. He has within himself the reason-view of all things possible to be put in objective being, in the plans or ideal archetypes to which they must conform, and his regard to that which is worthy of his own acceptance determines what of all that is possible shall also be actual. He is absolute liberty, for the one rule of that which is everlastingly worthy of himself, and securing his own dignity or glory, gives a repellency and exclusion of all ends that might tyrannize and enslave. He is absolutely blessed, for in his constant holiness and steadfast purpose, fixed upon his own glory, there is no collision or disturbance, but the perpetual serenity of an unruffled flow of righteousness. He is absolute sovereign, for while the ultimate end of his own dignity is ever before him and eternally directing all his agency, he as supreme has rightful authority and headship over all the beings that exist beside him, and may rightfully command in the ends of his glory that they should serve him with unquestioning and constant devotion. He is, in fine, and as the most comprehensive form of expression, THE ABSOLUTE GOOD; good in himself as supremely excellent without any reference to a further end, and good as the source and supplier of all the good that any other beings possess and enjoy. He can be put to no use as a means to get something beyond himself; but as the end of all ends, all other things fulfil their measure in conspiring to present that to him which is in honor of him. The highest seraph and the humblest saint honor themselves only in their devotion to his honor.

This conception of Deity as the Absolute Good, holds

itself completely above and beyond all conceptions that apply to nature, and are formed in the connections of the discursive understanding. Nature has its conceived powers, which in combination make up all the statics and dynamics of physical science; but within the Deity there can be conceived no action of physical forces, pushing, pulling, balancing, preponderating one with another. No combinations of substances can be here conceived, working their changes and making their qualities to pass away as one displaces another, which is the constant march of nature's causes and events; but the absolute Jehovah is in essence as in purpose, "without variableness or the shadow of turning." From nature, as giving a permanent position, we determine all bearings and distances, and thus judge all places to belong to the one space; and from the ongoing of nature we determine all succession and duration, and therefore judge all periods to belong to the one time. But in our conception of the absolute God we have no permanent points from whence to begin any measures of space, and no fixed instants from whence to begin any computations of time. In the Idea of the Absolute we can fix no "here" and no "there," no "above" nor "below," no "outside" nor "inside," no "shape" nor "distance." Just as little from this Idea can we determine a "now" or "then," a "before" or "after." Space and time are wholly irrelative; substance and cause have here no significancy. All these apply to nature, and the Absolute is utterly supernatural. He maintains his being without resting on any substance; he puts forth his agency without waiting on any cause.

As thus independent of nature, he can be conceived as

the Creator and Guide of nature. He has the conditions within himself for an activity that shall put nature in objective being, and fix the current of its flowing events to a channel that shall reach and consummate his pleasure. The moving spring to create, and to create thus and not otherwise, is in no constitutional want, no appetite he finds craving within him, but solely the conscious behest of what is due to himself and most worthy of his own accepting. It is love, in the acceptation of a pure pleasure in the right, and not the impulse to be happy; a regard to well-being and not merely to good-feeling.

His activity may thus originate, after the eternal archetypes which absolute reason supplies, a material universe wisely and exactly adapted to his design. Reason determines that the physical should minister to the sentient, and that the sentient should subserve the spiritual, and thus it may be anticipated that the creative work will arrange itself in ministering subserviences through the varied orders of material, vegetable, animal, and spiritual being, bringing out what was potentially within him, and guiding on to consummated issues, till the full claims of his own intrinsic excellency are satisfied. In such an Idea of the Absolute we have the conception of a God who is at once Creator and Governor; Author, Guide, and Finisher of a Cosmos, or glorious universe, whose only reason and end is that it is worthy of his own acceptance and approbation.

CHAPTER II.

THE ETERNAL PRINCIPLES OF THE UNIVERSE.

1. Matter is Force; distinguishable as Antagonist and Diremptive.—Spirit must be senior to matter. In the already attained Idea of an Absolute spirit, we have that which is essential in all spirit. That a spirit may be Absolute, he must have all the conditions and resources of independent action and direction within his own being, and thus stand wholly absolved from all limitation in any thing out of himself; but that a being should be spirit, absolute or dependent, he must have spontaneous activity or self-motion; an energizing which is initiative within his own being, and not a superinduced impulse from another. That such Spirit may be rational and free, he must find his own ideals in himself, and be competent to work from the archetypes and plans of his own origination. He must have his ultimate ideas of the beautiful, the true, and the good, and be able to criticize and judge of all beauty, truth, and goodness, by his own independent standard. Spiritual activity is always simple, without counteraction or reaction; there is no mechanical impulse, resistance, nor friction; no composition nor resolution.

But what is Matter? The first answer comes from the sense. The conception as gained in experience, the earliest, the easiest, and thus the common conception of matter, is that of some dead, dry, hard substance, given in mass of a larger or smaller volume. It is found divisible into parts, and we readily conceive that the largest bodies may be made up of small particles, and in our analysis of these particles we bring them into atoms which will not admit that we should further subdivide them. We thus begin with that into which we have analyzed our experience, and conceive of matter as originally existent in indivisible atoms, and that by various conjunctions of the atoms all bodies are formed. In all cases, whether as atoms or in the mass, matter is for the sense a lifeless, powerless, motionless substance; utterly inert, except as something is done to it, and in itself only existing to occupy and cumber the place it fills.

When matter is subjected to a nicer scrutiny in experiment, the conception is more extended, but not at all corrected. It is observed that matter in bodies is perpetually altering its qualities, and though often by slow gradations, yet in all cases, matter is moving from present modes of existence and transmuting itself to other forms. Liquids are congealed or pass off into vapor; fermentation carries its changes through the successive saccharine, vinous, and acetous stages; the atoms crumble and the hardest bodies become disintegrated, and these again are made the elements of new compounds; living agencies are assimilating and building up new bodies, and then the life goes out and the body again dissolves and their elements are scattered; and colors, densities, magnitudes, indeed the qualities of every sense, glide from one into another, and nothing abides permanently. All things flow. We sometimes speak as if one portion of matter moved or changed other portions, and that thus matter was conceived as itself active in producing its changes; but a partial reflection again qualifies the language, and we speak of powers and forces given to matter, and that the imparted force and not the dead matter does all the work and makes all the changes. The highest conceptions of the sense will therefore be, that matter itself is not cause except as a causal efficiency is given to it; that the forces and powers of nature are superinduced upon matter, and are something other than the matter; matter is mere *inertia*, and all changes are wrought in it and not by it.

But when such a conception is subjected to the insight of reason it is found utterly empty, and that nothing can be made of it but a mere negation. To attempt applying it to any use is an absurdity. What can this passive and inert existence *do?* At rest it cannot move, and moving it cannot rest, without a force supplied to it. It can neither change nor resist change, neither combine nor dissolve, neither sustain nor press, except as power is given to it to do all the work. Put it where we may it is utterly a *caput-mortuum*, neither acting nor reacting; the force given to it does all things for it, while the dead-head itself is incapable of any use, or of becoming a means to any end. How can it be *known?* If any sense receive an impression and thereby a sensation, out of which the intellectual action brings a distinct and definite perception, that impression and sensation must have been induced, not at all by the dead matter, but by some efficiency put into matter, and it must be this and not the matter that becomes object in perception. What then can it *be?* It cannot *exist*, for it cannot *stand out* in any sense; it cannot *subsist*, for it cannot *stand under* any quality; it is wholly a negation, and if we should attempt to conceive of it in any way as object, it would be the absurdity of an object that could not be put before any organ of sense.

We must, therefore, wholly renounce such a conception of matter, for indeed upon rational examination it will be found to be an impossible conception, a mere negation in the thought. Let us, however, keep this *force*, which we have supposed to be supplied to matter, and which we have found in such case must work all the mutations that occur in matter, carefully subjected to a rational insight, and determine whether indeed this force that does all that is done is not matter itself. Simple activity is spiritual activity, and has nothing in it that can awaken the thought of force; and it is only as it meets some opposing action and encounters an antagonist that we come to have the notion of force. In all push and pull there is counteraction, complex action, action and reaction, while simple spiritual agency can never be made a conception of physical existence. It cannot be thought as taking and holding any fixed position; it cannot become a permanent and have a "where" that it might be conceived to pull from, nor a "there" that it might be conceived to push to. It could not be determined to any time nor to any place, for it has no constant from whence the determination might begin nor where it might end. When, however, the conception is that of simple action in counteraction, an activity that works from opposite sides upon itself, we have in it at once the true notion of force. From the difficulty of clearly apprehending counteraction or antagonism in a single activity, as always acting in opposite directions upon or against itself, and which must be the true conception, for the notion is that of one source for the antagonism, it will be more readily taken and equally available in result, if we here, and generally through the work, conceive of two simple activities meeting each other and reciprocally holding back, or resting against, each other, and thus of the two making a third thing at the limit of meeting which is unlike to either. In neither of the two activities can there be the notion of force, but at the point of antagonism force is generated and one new thing comes from the synthesis of the two activities. To distinguish this from other forces hereafter found we call it *antagonist force.* In this, position is taken, and there is more than the idea of *being*, which the simple activities each have; there is being *standing out*, an EXISTENCE; being *in re*, *reality*, A THING.

Let, then, an indefinite number of such positions contiguous to each other be conceived as so taken and occupied, and a space will thereby be filled and holden; an aggregate force will maintain itself in a place; and a ground is given on which other things may rest. A substantial reality here exists. This antagonism may be conceived to be of any degree of intensity, and the substantial ground will hold its place with the same amount of persistency, and stand there permanent, impenetrable, and real. Nothing else may come into its place until it has itself been displaced. It is not *inertia*, but a *vis inertiæ;* a force resting against itself, and thus holding itself in place. It *rests*, because it has intrinsically an equilibrating *resistance.*

But this conception of antagonism alone, though fully adequate to give substantial matter, will not be found adequate to give such forms and modes of matter as a universe needs for the rational ends designed in it. There will need to be varied substance; combinations and resolutions; perpetual changes and processes through successive stages; and thus our very primitive idea of matter must comprehend more than the idea of pure antagonist force, even that which may dissolve and become a combination with pure antagonism. We conceive then of an activity going out in exactly the reverse process of our antagonism, even a beginning in the same limit of the meeting simple activities and working on each side away from the limit; a throwing of simple activities in opposite directions from the limit of contact. Not a counteracting and resisting, but a divellent and disparting activity; not an antagonistic, but hereafter known as distinctively a *diremptive* movement. Such an activity could not be conceived as space-filling of itself. Wherever the limit in which there might be conceived the contact of two simple activities should be, the diremptive movement would be away from that limit on each side, and thus a space-vacating and not a space-filling activity. The diremptive movement alone would be a disparting and going away of the activities from each other, and leaving a void. But if this diremptive movement be conceived as at tne very limit and point of contact of the antagonism, the antagonist activity working toward itself in the limit, and the diremptive activity working from itself out of the limit, then must the diremptive movement on each side encounter the antagonist movement, and the simple diremptive activity going out on one side from the limit will meet the simple antagonist activity on the same side coming in to the limit, and these two simples of the opposite kinds of forces must make a new counteraction among themselves. And equally so with the going out and the coming in of the opposite kinds of forces in their simple activities on the other side of the limit, the one must encounter the other and engender a new counteraction among themselves on this other side. The result thus must be that while the diremptive activity disparts and loosens the antagonism, the antagonist activity on the other hand restrains and binds in the divellency, and thus the diremption can neither go off wholly on either side and leave the limit void, nor the antagonism come up from each side and make the limit full, but both antagonism and diremption meet in the limit and make a third thing, which may be called indifferently an antagonist force loosed, or a diremptive force fixed.

The pure forces in their contact in the simple limit may be known as units under the term of *molecules*, or molecular forces; the working to the limit constituting an antagonist molecular force, and the working away from the limit constituting a diremptive molecular force. The combination of these forces, in their joint interaction making a new compound as a third thing unlike to either alone, may be known as also a unit, constituting a material *atom*, and which may further on be known as a *chemical* atom or molecule. Our conception of matter must therefore be of this combination of distinguishable forces, though we shall find it convenient for the more clear apprehension of the principles of the universe to follow out the workings of each distinctly and separately.

2. Creation.—In the manner here used creation has

the meaning of origination; the putting forth of something where before there was nothing, and this something thus set forth a new thing which had not previously an existence. It does not involve the impossible thought of existence coming out of a void of all being. The axiom, "out of nothing, nothing comes," is to be taken as universally conclusive. There can be no creation without a Creator; and as the creation we now seek to apprehend is that of the beginning of existence, or outer being, the Creator must himself be conceived as the uncreated; he who ever *is*, and yet who never *exists*. His being is never objective; expressed in form; standing out in definite proportions; but purely spiritual, and known only in that supernatural light to which no mortal can approach. This ever invisible Creator puts forth a material creation in objective palpable manifestation, which only in his putting forth began its existence. How shall we attain to a rational Idea of such creation? The intention here is simply to attain the conception of originated existence, leaving the detail of the completed genesis of the universe to many subsequent separate paragraphs.

With the distinct conception of force as the essence of all material being clearly in mind, we seek now to apprehend how, where force is not, it may begin to be. Force cannot come from utter emptiness. Nor is it now to be apprehended as produced from some antecedent force, and thus a propagation or production from some force already created. Forces may change their modes of manifestation indefinitely, and this will be but the progressive development or successive births and growths in nature itself; but we now want the conception of nature's origin. The great difficulty to a clear apprehension is removed by keeping a steady discrimination between the functions and cognitions in the discursive understanding and those in the comprehensive reason. To the understanding nothing can be conceived as literally beginning to be. That which is must have been a production or outgrowth from that which before was, and all new things are only some changes in the modes of old things. The whole function of the understanding is to connect in judgments, and the subject must already be given in which to connect the new predicate. The predicate must be thought as already belonging to the subject. If there is a new quality, it cannot be viewed as then a thing newly originated, but a change in some substance that had before existed; and if there is a new event, it cannot be thought as then having its origin, but only as coming out from some old cause in a new mode of existence. The water following the ice as dissolved is not new, nor the vapor as following the water, but the new predicate is ever from some old subject. For any thing absolutely to begin to be, in the discursive judgments of the understanding, would be the absurdity of attempting to connect with no medium; of thinking in judgments with no subject for the predicate. To this faculty, that which is must ever be the product of something that before was; a change of some old existence into a new mode of manifestation. Nature thus, from first to last, goes onward the same, with no originations nor annihilations, but only a perpetual passing of the same substances into new modifications.

But when we keep the reason-idea of spiritual agency, as spontaneous activity self-directed, we shall have an utterly new kind of cause, viz.: a cause originating or causing to be from itself, and not a cause conditioned or caused to cause from something back of itself. It is activity in liberty, which can make a beginning from conditions within its own being. We have in this conception no impossibilities, nor absurdities of the *last-first*, in affirming that we may intelligently apprehend how an utterly new thing can absolutely begin existence. With all rational spirits there is such capacity of initial causality, and thus of all free and responsible beings, we affirm that their personal acts are their own origination, and can no more be transferred to any other person than their separate identity. Man and angel can, in this sense, truly create. Their good or bad deeds are of their own origination. Whatever another agent may do in throwing his own conditions upon them, he does not originate their acts within them.

But man cannot originate new forces, and thus man cannot create matter. He is himself incarnate; utterly merged in matter; and can thus put out no act that shall immediately meet another act in counteraction, but his every act of energizing must first encounter the forces in which he is incorporated. His activity meets forces, and moves matter already created, but his activity cannot, with nothing between, meet itself in counteraction, and take a new position, and thus begin a new space-filling operation. Yea, if we were to conceive of angels as pure spirits, activities without corporeity, and thus competent to make one act counteract and hold another in position, yet these counter activities could only be within their own subjective spheres, and condition their own conscious activities, and could be no forces to condition other agencies which could not bring themselves within their subjective

spheres. Thus, neither man nor angel can be conceived as competent to create force that shall be objective, real, substantial, and impenetrable to another agency.

But with the conception of a Supreme Absolute Spirit, all these difficulties are excluded. He can begin action, he can put action in counteragency with no forces intervening, and whatever positions he may thus take and hold by permanent forces, though subjective to himself, or within his own sphere of agency, they may be objective to all other being, for all being will be alike subjective to Him in whom all live and move and have their being. Take then the Idea of the Absolute, already attained, and within the pure spiritual agency of his being there is no force; no antagonism or counteragency. Simple spiritual activity takes no positions, fills no space, puts within itself no limits from whence we can begin to determine places and periods. Spaces and times are here wholly irrelevant, and as there is no fixing in place and moving in successions, so nothing of impenetrable substances and series of physical causes can be thought as lying and working on in the Godhead. But in the knowledge he has of his own supreme excellency of being, there is an end in his own dignity and glory ever before him. He knows what is due to himself, and nothing can intervene that he should not be true to himself. "He remaineth faithful, he cannot deny himself." He sees that it behooves him, as a right consciously due to himself, to manifest himself in creation. Under such ethical behest, and not at all before the impulse of any constitutional craving, God arises to the work of creation, and becomes a beginner and Author of an existence which before was not.

Solely from the reason, and not from any want as if he too had a nature, God puts his simple activity in counter agency. He makes act meet and hold act, and in this originates an antagonism which constitutes force; a new thing; a something standing out for objective manifestation, and holding itself in position as a reality distinct from his own subjective simplicity. This force fixes itself in position; holds itself at rest; and so far from being inert, its very existence is a *vis inertiæ*, or a force actively holding itself still. Combined with this antagonist activity, in the same limit of counteraction, is the diremptive activity that works conversely to the antagonism, and which though hereafter to be considered distinctly, may now for the present be apprehended as in unity, and the antagonism and diremption to be the one agency of the Absolute Spirit in one and the same limit of their action; the antagonism working each way into the limit and the diremption working each way out from the limit, and both making in their interaction a compound material substance, which has the *disparting of the antagonism* in the diremption between the counterworking activities, and the *fixing of the diremption* by the antagonism on each side of the divellent activities. There is thus the combination of three molecular forces in one limit—the diremption works each way out from the limit, and thus counterworks with an antagonism coming up each way in the limit, and thereby two antagonisms and one diremptive force equilibrate each other and fill and hold the space they have taken. Any considerable extent of space so filled, a cubic inch, or a cubic mile, is a creation of matter palpable to the senses, impenetrable and substantial. The force in every position

becomes a force reciprocally acting through the whole place filled, and is at once subject to its own inherent laws and bound on its course of necessitated successive development. It is a *nature*, having already in it a conditioned and predetermined series of growths which must come out in their own order.

The simplicity of the spiritual works on still undisturbed within the Deity, for no conditions of the material reach back of the point of counter-agency. In matter is force, or the physical, and all its necessitated efficiencies work downward in their destined sequences, but above matter all is still spiritual, supernatural, the free ongoings of spontaneous activity directed upon the end of its own dignity or glory. The physical cannot push itself back and hinder, tire, or in any way interrupt the activities of the spiritual; and the spiritual cannot bring itself down, and reveal its agency amid the statics of the substantial or the dynamics of the causal, and thus appear on the same theatre with the physical phenomena, but the natural and the supernatural spheres are forever separated in the limit where simple agencies in counteraction become a new thing that holds its place, and works its way, as a physical force necessitating its products. The creation of the material is from God; its genesis is in him; its perpetuation and sustentation is from the continual going out of his simple activity; but this material is not God, nor at all competent to rise from its imposed conditions into the place of the Absolute. The Logos, or divine working word, is *in* the world; is the life and light of the world; and yet he was in the beginning with God, and ever is God, while the world is not he but his creature.

3. The Determination of Space and Time.—There are many kinds of space and of time, and one may appertain to one person and another to another person, and that which has pertinency to one may be wholly impertinent and irrelevant to the other, and if left to their own sensible experience, or the deductions of the logical understanding, neither one could determine his own spaces and times to belong to one space and one time, nor that the spaces and times of both belonged to one common space and one common time. These very significant propositions are usually wholly overlooked, and most persons make no question of any distinctions in kind, in spaces and times, and thus apprehend nothing of the difficulty or the importance of the determination of spaces and times. A few obvious but disregarded facts presented carefully to the insight of reason will very fully convince us that our experience, or our logical judgments, have nothing to do in determining space and time, and that such determination can only be made by fixed forces, in position, and their perpetual changes in succession.

There may be a morbid affection or some unnatural distortion given to the eye, which shall induce fantastic colored spots even when the eye is closed. With such phantasms, a space is also given in which they appear, but so soon as the colored spots vanish, the space in which they were goes out with them. If on another day other such phenomena occur, a space again is given in which they have their different places. Now neither the experience nor the judgment can determine that these spaces are the same, or that they belong to one and the same whole of all space. They came with their colored spots and went with them, and each space is in the experience as distinct from the other as the spots in one are distinct from those in the other. Two persons may have such phenomena, but the spots and the spaces are for each person his own, and one can no more say the space he has is the same that the other has, than that his spots are the same. There are here many different spaces.

So when I construct any pure diagram, as a triangle or a circle, that pure figure has a space in which it is given, and when the figure drops from the inner consciousness, the space in which it was is lost also. I may successively make and lose many such constructions, but I shall always have and lose their spaces with them. Two persons, or any number of persons, may be constructing their pure figures, and they will all have their spaces with their diagrams, but neither the one can determine all his spaces to belong to one whole space, nor the whole number of persons determine that all their spaces belong to one space in common. Much less could any determine that the spaces in the affected organ, as before, and those in the inner consciousness, as now, belonged to one space.

Thus, again, of any mirror; there is a space whenever there is a reflected image; but if the mirror reflect no image, it will give no space. The spaces come and go with the reflections. A cloud passes over the face of the lake which had mirrored the heavens and the objects on its borders, and the space and the images go together, and when the cloud has gone the images and their space again come. No experience nor discursive understanding can make the different mirrored-spaces stand together in one space, nor all spaces into one whole of all space.

Just so also of time. I may be absorbed in mental reflection and take no note of outer successions, and I shall be conscious that a time has been passing in that my inward thoughts and reflections have been succeeding each other. I may again arouse myself to a consciousness of outer objective successions or may have an interval of sleep, and afterwards another experience of reflection or musing meditation may occur, with its consciousness that a time is passing though I take no notice of any outward ongoing. Here will be different musing-times, but no experience nor logical judgment can affirm that they belong to one time. Two persons may so meditate in separate absorption of all outer consciousness, and each will have his own musing-time because each will have his own inward successions, but neither nor both together can put their distinct times into one time.

Or, again, one may dream, and after an interval of deep sleep or waking consciousness may dream again, and with each dream there will be its own dreaming-time, but no experience or discursive thought can put the different dreaming-times into one time. Two or more persons may so dream, and each have his own time in his dream, but no one nor all together can connect their separate dreaming-times into one common time for all. How much less shall any one make all musing-times and dreaming-times to connect in one common time. The times are given in reference to the successions which pass in the inward consciousness, and if there are alternations of conscious movement with suspensions of all movement in unconsciousness, the interrupted experience can in no way join itself into one, nor can any connections of the logical judgment bridge over the chasms.

Even so with all *objective* spaces and times; the experience and the logical thought can never make them to be in one space and one time. I look upon some broad landscape and determine all its distinct objects relatively to each other, and make the whole to belong together in one place, as a space which contains them. But if I am removed in my sleep to another position, and I awake again and look upon another landscape, I can again make all its objects to belong to one place as a space containing them; but I can neither by experience nor judgment put these two spaces together into one space, and say which direction the one is from the other, nor determine that the two belong indeed to the one whole of all space, for neither my experience, nor my logical understanding, as an induction from all that experience gives, can determine that there is any one whole of all space. And two or more persons may each have their landscapes with their different objects determined in their relative positions in one place as a space containing each landscape, but neither one nor all these persons could, from their experience or their logical thinking, determine that the spaces which held all their landscapes, respectively, belonged to one space, nor even that there was any one space which contained all spaces.

And so also I experience a series of successive changes in surrounding objects, as the passing of different shadows and changes of color over the landscape, and I can determine them in their relative periods in the one time of duration for them all. And if I am in unconsciousness stopped from all experience, and again watch the changes of a landscape in some other conscious experience, I can put all the occurrences again into their relative periods during the one time of their successions, but if this last also be cut off from all succession in unconsciousness, I cannot say, either from my experience or my logical thinking, that these two times of landscape-changes are in one time, or which is before and which after the other. The chasms of unconsciousness sunder the continuance of successions, and when the experience has been cut off from both completely, they stand each in their own time, and nothing is given to permit the connecting of the two times into one and determining their order of occurrence. And if two or more persons had their conscious objective successions, they would each have their times, and their determined relative occurrences in their times respectively, but neither one nor all could put their respective times into one, nor show any relative order of occurrence in reference to them.

Each man's spaces and times are only his own spaces and times, and where they have been disjoined in his experience, his discursive thought can never put them together, and much less can any one man put all the spaces and times of all men into one space and one time. So it must ever be, when all men are left only to experience or the deductions of the logical understanding to determine space and time; they can never bring their own distinct spaces and times into one space and time, nor ever possess one common space and one common time between them. Experience gives them many different spaces and times, and no judgment from experience could put the different spaces into one space nor the different times into one time. Each man's places and periods would be for him just as he constructed them, and he could determine nothing for another.

But the insight of the reason finds at once, that if there be one substantial existing object that holds its permanent position for the same man through all his subjective states of consciousness, this will enable him to determine all places from this one place; and if this one substantial object be common to all men, it will enable all men to determine for themselves one common space. The one man, and all men, can construct bearings and distances from one and the same position, and thus each man and all men may determine one and the same space. And also, if this substantial object vary its phenomena successively, it will enable each man and all men to come to the same successions, and determine one time for each man's times and all men's times. It is from this necessary principle of space and time-determinations that in Rational Psychology we demonstrate the existence of a real substantial universe against all Sensationalism and Idealism. All men have the determination of one common space and one common time, but this could not possibly be, except upon their communion with the same substantial nature both in its permanence and orderly successions.

We thus know that when as yet nature had not been put out in objective manifestation, nature's place and time could have no determinate significancy. If the present nature of things be annihilated, the determinate places and periods of nature would vanish with nature itself, and if there were minds to have inward experiences, they might construct inward figures and have subjective spaces, and limit successive movements and have subjective periods;

but each mind only for itself, and could never put each his own places and periods into one space and one time, nor bring all men's spaces to one common space, nor all men's times to one common time. And should another nature of things be created, it would come up in its own determinate space and time, and no understanding could connect it in the same one space and one time of the nature which had been annihilated. All minds must be able to go to one and the same substantial nature of things for the determination of their spaces and times, or they can never determine that they have one common space and one common time.

When, therefore, we conceive of an Absolute Creator, setting his simple activity in counteragency and taking a position, and in balanced antagonism holding that position permanently, we have in it all that is conditional for space and time determination. In the simple spiritual agency we can determine nothing of space or time, for there is no fixed point from whence to determine direction and distance, and no fixed instant from whence to determine successions and durations. But in the genesis of a force there is a determinate place taken, and in the developed progress of its working according to its inherent conditions there is a determinate succession, and each man may determine here his own places by the one space, and his own times by the one time, and all men may here determine one space and one time in common for them. While each man's subjective spaces and times are still his own, and no other mind can come in communion with them, the one substantial space-filling and time-enduring force is common to all for the same space and time determination.

4. MATTER MUST IMPRESS ITSELF UPON THE SENSES.—If matter were conceived as wholly inert, it would be utterly inexplicable how it should affect any sense, and its qualities become perceived through any organ. Whatever was done must be by some imparted force, and that would give the whole impression, while the dead matter would be entirely superfluous. As it could affect nothing so the senses could perceive nothing of it. But when we conceive of the substantial matter itself as a force filling certain places in space, we may readily apprehend how the senses must be impressed by it, and the sensations induced be brought up into the light of consciousness.

The sense can never go back of the sensation and determine any thing of the substantial being which gives the sensation, for the sensation in the organ is the only content or material out of which the perception is made. The substance affects the organ, and the intellect distinguishes and defines this affection in the organ, and this only appears, or becomes a phenomenon. We can never perceive the substance, and only the peculiar manner in which the substance has affected the organ, and hence the sense can only give us the qualities of things and not the things in themselves.

But the insight of reason penetrates the act of perception itself, and comprehends the sense in its complete function, thoroughly. The organ must in some way be affected from without itself, and this affection induces the sensation within itself, and as this has its own peculiar modification according to the nature of the organ affected, so that must be *distinguished* by the intellectual agency and the peculiar quality determined, and this determined quality must, by a still further intellectual action, be wholly *defined* and the quantity of the quality fully perceived, whether in space, time, or amount. The qualities perceived are but the modes in which the substantial reality impresses itself upon, and thus manifests itself in the sense. And with this conception of a space-filling force, it is quite competent for the understanding to trace its necessary connections in any organ, and thus attain the clear idea of the functions of the sense through all its kinds and varieties of perception. The inherent energy in matter itself is sufficient to impress itself upon the senses, and make its qualities to be perceived. We will go through the senses in the order that the impressions will be most readily apprehended.

The Touch.—The essential being of matter is force, or counteragency, and it is the nature of this, in its balanced action, to hold itself permanently in the position it takes. The organ of touch, here anticipated as already existing, is the finger or some fleshy part of the body, and is thus itself, like all matter, a composition of space-filling forces which hold themselves permanent in their place, and moreover the finger possesses a vital and sentient activity which penetrates every part, and capacitates it for communion with the intellectual agency which must distinguish and define its content. The organ, therefore, has its own place, and its conditioned nature to retain its position, and its medium of communication with the intellectual and conscious spirit.

When this organ, then, is made to meet any matter, either by invading the place of another portion, or by being invaded itself by another, there must supervene in the contact a reciprocal pressure, and which may be more nicely regulated by the voluntary action of the living muscles, and by this impression an affection or proper sensation is given to the vitalized material organ. The matter touched may be made solid by any different degrees of intensity in its constituting forces, and may also have its surface of any variety of shape and outline, and which must determine the accordant impression upon the organ; and when this is distinguished and defined by the intellectual action, the perception of the quality, as hardness, roughness, weight, &c., will be perfected. This impenetrable space-filling force may impress the sense of touch with any conceivable degree of resistance, and must thus give to itself in the perception the like degree of intensity as quality.

The impression is, in this sense, made only by actual contact, and thus no condition is given whereby to determine distance from the organ; but relative distances, as extension in space and outline of figure, may readily be determined by a continuous application of the organ of touch to the resisting matter. So far as there is a continuous progressive contact, either by the organ moving over the matter or the matter moving on the organ, the occasion for determining extension, shape, and size is given, and the sense can thus perceive all the qualities of length, breadth, thickness, and complete shape which the matter may possess.

In all the above cases a conscious measure of muscular pressure has been necessary to the perception, but a much lighter contact with the opposing matter in a slight friction upon it, will give phenomena that are considered rather as sensations awakened in ourselves than as qualities possessed by the matter, and it is only in such a degree of intensity that the matter occasioning it forces itself upon our attention, that we can refer the phenomenon as quality to the matter itself. Thus with the feeling of irritation, titulation, and gentle warmth or coolness, when slight, we say *we* feel the sensations; but when more intense, and the thought of the matter occasioning it obtrudes itself, we say the *body* feels rough, harsh, hot or cold, and we apply the sensation at once as qualities of the matter.

It is herein manifest that nothing further is necessary to conceive, as the inherent essence and constitution of matter, than a force filling and holding its position in space, and all the possible qualities which the touch can perceive must be very intelligently occasioned by it.

The Taste.—All matter in a mass, larger or smaller, is a compound of the elemental forces each in its own position, and thus each point of force may be taken to be a molecule of matter. These must also be more or less intense or must have varied directions and combinations given to their antagonism, and thus the molecules of matter must be of great variety. The atomic existence and varied composition may give occasion to all the varied forms of matter in what is known of earths, metals, salts, alkalis, acids, &c.

The tongue, also, as the organ of taste, with the surrounding parts of the mouth, is here anticipated as having its composition of corpuscles, and the whole vivified and sentient with that living activity which has assimilated and incorporated them. This organ, from its own construction, is hereby fitted to take on the peculiar sensations given when the sapid matter is in contact with it. The conditions for the sensation of taste have this peculiarity in distinction from the touch, that there must be not merely contact, but a dissolving of the compound body, and the bringing of the separate molecules upon the organ in their own particular degree and variety of pungency. These different particles give occasion to different sensations, and when these are intellectually distinguished and defined there will be all the conditions for complete perception.

The Smell.—The living assimilating process builds up also an organ, for attaining the different odors by which the material world is qualified, in the nose. Penetrated as it is with sentient life, it becomes competent to receive the impressions which the effluvia from surrounding bodies may make upon it.

The essential elements of matter in the primitive forces which compose it may be more or less intensely held in position, and more or less firmly adherent, and thus some portions of matter may be permanently adhesive and must therefore be inodorous; other matter may be volatile and admit that the various mechanical agencies that surround it may readily force off some of the elemental particles and thus effectually surround itself with an effluvia from its own substance. The tension of the particles, and the energy and direction with which they must be thrown upon the organ will determine the impressions made, and thus from either the conditions in the effluvia itself or those of its transmission to the organ, there must be sensations of all varieties, giving occasion for distinctly and definitely perceiving all kinds of odors. The aroma may so stimulate and excite as to awaken the most regaling fragrance, or an effluvia may be present that shall give the most fœtid and offensive smells.

The organ is not in this sense necessarily brought in

contact with the body of matter itself, as in touch and taste, but only in contact with some of the particles sent off from the body, and thus there can be no opportunity given for constructing shape and outline by the smell, and only the capacity for vaguely estimating distance and direction from the degrees of intensity with which these particles may strike upon and impress the sense. A sufficient condition is however here supplied for all the perceptions that can be gained through the sense of smell.

Sound.—With a sentient organ like the ear, there is a capability to receive impressions from material nature of an entirely different kind than those in touch, taste, or smell. The organ itself is expected to stand wholly and often quite distantly separated from the sonorous body, and no part of the material substance is itself to flow off and meet the organ. A medium must be supplied in the elastic space-filling force which constitutes the air, and which surrounds both the organ and the sonorous body, and fills the whole space between them. An ear in *vacuo* must be without sound. The sonorous point is at the ear, but the condition given for the sound may be at a point very far removed from the ear, and the impulse from the body which puts the elastic medium in undulation must make the communication from the body to the organ, and thus all the conditions for the appropriate impression and sensation are supplied.

The stroke which starts the impulse and the solidity of the body stricken, and the rarer or denser medium through which the communication is made, must all modify the sensation and determine the variety of the sound. It must thus be louder or weaker, and modified through manifold tones which have a higher or lower pitch. The impulse at the starting point in the sonorous body must perpetuate itself through all the media, and the impression must be determined by all the peculiarities given in the whole process. In this way an occasion is manifest for all possible sounds through the matter made of compounded space-filling forces. And not merely the direct impulses, but the rebounding and reflected waves from some intervening sonorous body, will occasion all that is to determine the perception of the echoes which may accompany some original sounds. All the laws of acoustics are intelligently read in the nature of these space-filling forces.

Vision.—With the complicated and nicely adapted organism of the eye given in conception, it may be a clear insight of the reason that matter, as a space-filling force, must give all the conditions necessary for vision.

Like the ear, the eye also is adapted to receive impressions, not directly from contact with the object, but through the medium of that which may lie between the organ and the object. This intervening medium may be a direct transmission from the object, or some force that shall put in oscillation an elastic fluid lying between the organ and object. What this force is, will hereafter be determined, but it is sufficient here to have the conception of a space, about an organ of vision and an object, filled up with contiguous antagonisms, and which may be put in motion and made to affect the organ as the movement shall be modified by the object. The medium must in this way give its impression to the organ from the object, and this impression, whether by linear or oscillatory impulse, must be perpetuated to the retina, and through the optic nerve to the sensorium, and this must give all the conditions that any conceived matter may present for distinct and definite perception.

A more full apprehension of the force which is to make bodies luminous, will give a more complete and adequate insight into the necessary determinations of vision, but enough is given in the idea of a space-filling force, to apprehend that this, and not any dead matter, must be the medium of perception by sight. The force which must move such inert matter would be all that could impress the organ, and when we have the force given in idea, the dead matter may be altogether dispensed with as wholly useless and irrelevant.

Thus it is that the whole origination and endless modifications of our phenomenal experience, which is communicated through our organs of sense, have their sufficient conditions in the one idea of a space-filling force, which may vary itself in intensity, rapidity, and direction of agency, indefinitely. This substantial matter must make its impressions upon the organs according to its conditioning nature and their organic constitution, and must thus reveal itself through them in determinate modes. The forces constitute the substantial existence, the modes of organic impression and perception constitute the phenomenal or qualitative existence.

5. Statics and Dynamics.—The sense-conception of matter can by no possibility admit of any thing static or dynamic in nature. The supposed matter is wholly dead; mere *inertia;* and can possess nothing by which it may be conceived as holding itself in place whereby it may sustain any thing, nor as moving from its place whereby it might

push or pull any thing. If it upheld any thing, as a static, it must itself be sustained by some *ab extra* force, and if it repelled any thing as a dynamic, it must itself be pressed by a force not its own. The forces introduced, to determine the dead matter, would do all the work without the superfluous introduction of such an inert extension.

But our thought-conception of a space-filling force as the true substantial matter involves the full conception of both statics and dynamics. Counteraction in equilibrium must stand self-fixed. It is a force holding itself in its place. It is competent to sustain a pressure equal to the energy of its own antagonism, and can be displaced only when the intruding body shall carry with it the energy of a more intense antagonism. Thus with all masses of matter. In any form or magnitude, there is necessarily a point towards which all the outlying points tend, and on the force in which all the forces in the whole mass are sustained. When that point is at rest all the other points are held at rest by it, and when its *vis inertiæ* is overcome, all the forces in the other points of the mass will also yield their *vis inertiæ*. A static force is that antagonism which holds itself at rest in its balanced counteraction.

A dynamic force goes to the overcoming of a static. It may draw or expel, but it goes to the removing another force at rest, or to the retarding or accelerating another force in motion. Should the dynamic not be sufficient to overcome the static, still, in so far as its intensity of antagonism goes toward this, it is thus far dynamic though the static does not yield to it. And this is involved in the very conception of matter as a space-filling force. The counteraction in any point is static when in equilibrium, but when the agency in one direction is more energetic than that in the other, though there is still counteraction, yet must the weaker yield to the more energetic and the whole counter-agency perpetually displace itself, in the direction of the working of the more strenuous agency. Such perpetually moving space-filling force may impinge upon another force and impel, or may be attached to another force and draw, and thus in any direction, the antagonist force in motion is a dynamic, either of impulsion or tension. A dynamic either drives or draws.

It is also obvious that a static is nothing in nature without a dynamic, for were there no push nor pull there could be no holding of place by an equal antagonism; and so also that there can be no dynamic in nature that has not also its static, for no push nor pull could be without a standpoint. In nature, there is a complete sophism of the ὕστερον πρότερον; and were there no way of attaining to the supernatural, both the perpetuation of rest and the beginning of motion would be absurdities; for you must first have your motion in the very act of *holding* at rest, and you must first have your rest as the hold-point or springboard of your *moving* some other body. The only way out of such an antinomy, between nature in the understanding and nature in the sense, is the apprehension of a supernatural in the reason. An absolute spirit has the spring to an originating act in himself, in that he is ethical law in his spiritual excellency to govern himself. He may originate action, directly from the claims as known to be due from himself to himself. He has an *ethical* stand-point and spring-board, and can thus put forth his spiritual act in counteraction and make a beginning. Spiritual activity put in counter-agency makes a *physical* stand-point; takes a position and holds it; and in that a static force already is, from which all physical mechanics may go out in operation.

6. PRINCIPLES OF MOTION.—If the mind be filled only with the sense-conception of matter as mere *inertia*, then can there be no apprehension of any principles of motion, and all its laws must be arbitrarily imposed. The very laws are mere facts, and for aught we can know, they might have been any other way as well as the present. With such a conception of matter, it was a toilsome and tedious process to find how in fact matter had been made to move, and then generalize the facts as far as possible and call them *laws* of motion. Dead matter cannot move any way, nor energize at all, and thus no thinking about matter, no insight of it, can get any motion, much less any laws of motion, out of it. The mind can only learn by experience how matter does move, and then generalize these facts and say they are *laws* of motion. In the true substantial matter, as space-filling force, the eternal *principles* of motion are already given, and may be found.

We will take the laws of motion as experimentally attained, and we may successively see, that the immutable principles of space-filling forces will necessarily determine every law of motion to be as it has been found that it is. It is thus truly a principle in matter, and not merely a law arbitrarily constituted.

The *first* Principle of motion is, *that it must be rectilineal and uniform.*—This is a necessary determination of the reason in its insight into the grounds of force from whence all locomotion must be generated. When two simple agencies counterwork, the result must be a resting against each other in static equilibrium, if the countervailing activities are of equal energies. If one activity be of greater energy, it will be counteracted by the other to the amount of its energy, but the excess of energy in the former having nothing to balance it will forbid that it should be holden in any one point; and yet, as the weaker activity continues its antagonism to the amount of its energy, there is a perpetual space-filling force, and which cannot be holden in any one point of space. The result must be a constant force which cannot abide in any one position, and is thus the idea of the generation of motion. A space-filling force, which cannot continue in any one point of space, is a space-filling force successively occupying different spaces, and is thus matter moving. Let these activities continue their respective energies in counterworking unchanged, and the force which balances the weaker energy will make its essential matter to be a permanent existence, but the excess of energy will make this permanent matter to be perpetually changing its place. The motion must be *incessant.* But this motion is generated only in the excess of the greater energy, and that is perpetually in its one line of antagonism to the weaker activity, and must, therefore, determine the motion to be in its own invariable direction. The motion must be *rectilineal.* And this excess of energy in the greater over the less is invariable in degree, which must secure the passing from point to point to be in all points at the same rate. The motion must be *uniform.*

This will be true not merely of one point of space-filling force, but must hold invariably true of any aggregate amount of space-filling forces in a body. If all the points of force be invariable in their comparative energies, the one principle must include all. And they must all move incessantly, rectilineally, and uniformly.

And this principle must equally determine the transmission of motion by impulse. If any amount of space-filling forces occupy their places at rest in their balanced action, and other forces moving come in contact in the line of their balanced antagonisms, the moving forces just bring their excess of energies in their direction of motion to the forces at rest, and add this excess to the activities working in the forces at rest in the same direction, and thus make them to have the same disparity of energy and in the same line of working, and thus the forces at rest must take on the same incessant motion, and in the same right-lined direction and uniform progression. And if the forces moving come in contact with other forces moving, by reason of greater velocity, the excess of energy on the one side of the antagonism in the swifter body will add its greater degree to the excess in the slower body that is less than its own, and this must quicken its motion; but henceforth that quickened motion must be incessant, right-lined, and uniform. The constant excess in its own direction must ever determine to a rectilineal and uniform motion.

The *second* Principle of Motion will find its expression in the following formula—*that motion which any superinduced force would give must be compounded with the motion which the forces already have.* This will apply universally, and introduce a new principle beyond that which determined the motion in the former case. The principle in the first case was, that the more energetic activity must move the unbalanced force directly and uniformly in its own line. There is now to be a combination of forces, and there must therefore be a principle modifying both the old uniformity and the direction. Another degree of excess in the antagonisms is given, and the old uniformity cannot continue; also an activity transverse to the old antagonism is contemplated, and there cannot be the rectilineal movement before the greater energy. Both the degrees and the directions of the forces must be compounded.

We take any matter moving under the control of the first principle of uniformity, in the line of the excess of the antagonist activity, and now superinduce a new force. It may be applied in the following directions and degrees precisely *in the line* of the old antagonisms. It may be in the direction of the weaker energy of the moving forces, and yet not sufficient to balance the excess of energy in the stronger; and it is then clear in the insight of reason, that it must retard the movement by just the degree of energy added to the resistance of this weaker side of the antagonism. If sufficient to just equal and balance the excess, it must wholly suspend all motion. If sufficient to give to the weaker side of the antagonism a stronger activity, the excess of energy changes sides and the old motion is not merely suspended but must be directly retrograde. If the superinduction be on the side of the more energetic activity, there must be an acceleration to just the degree in which the old excess of energy has been augmented. In all the above cases, it is manifest that the old motion is to be compounded with the new motion given, inasmuch as these compound motions are the resultants necessarily of the combining of the old and new forces, and thereby modifying the excess of energy which generates the motion, though there can in these cases be no change of direction.

But the superinduced force may also be applied *transversely* to the old antagonism. In such case, there can be no balancing of the antagonism, nor direct reversal of the excess; no merely increasing of the weaker nor the stronger activities, and thus no compounding of the forces and their movements can have any thing to do with the merely uniform rate of movement, but will necessarily modify the direction, inasmuch as this new transverse force will not admit of the old excess of energy to go any way up or down its old line of working. This old excess of energy will continue in its old direction, and the superinduced force will come and continue in some transverse direction, and the first principle of motion can have no unhindered application. The movement cannot be in the line of the old more energetic antagonism, for the superinduced force now thwarts this by cutting across its line; and no more can the movement be in the line of the new force, because the old excess of energy continues to work in its former direction, and must thwart the superinduced force.

This new force may come in any direction on either side of the line of the old antagonisms, but in any way, it must be in the same place with the old activities, and meet them in their common point of counteraction. That superinduced force is thus a third activity, meeting the antagonist activities in their point of contact, and interfering in the results of their working, and the motion induced must be determined by the compounding of all these activities. The excess of the antagonist energy, and thus the motion, was before on one side and in one direction of the antagonism, and the new force tends to move in its own direction, and they can now only neutralize and balance themselves in some common point between them. That common point will give its excess of energy as a unit, and move the force or molecule of matter accordingly, and the perpetuation of the activities must perpetuate the points in which they balance each other, and the motion must be through these points successively from one to another, and thus the line of motion must be through the points in which the compound agencies balance each other.

The rate of movement, and the direction which the excess of energy on one side of the antagonism has engendered, being given, and then the rate of movement, and direction which must be engendered in the excess of energy on one side of the force to be superinduced, being known, we must compound the two after their respective ratios and directions, and that must be both the direction and velocity of the newly-acquired movement. Geometrically, it is manifest, this compounding of the excess of energies in the two forces must give its line between their directions, and dividing the angle their lines of direction may make. If of equal excess of energy, and moving at right angles to each other, their compound must be a bisection of the right angle between them; and if of equal excess of energy but moving in direct antagonism, their composition must be in a line perpendicular to their common line of antagonism. If of unequal excess of energies their composition must give the line dividing their angle in the inverse ratio of the excess of energy, viz., the greater excess to have proportionally the less space, and the less excess to have proportionally the greater space, on their respective sides of the divided angle between them.

This principle of compounding the motions of two forces, which are generated by their respective excess of energies on one side of their antagonist activities, is applicable to any number of superinduced forces, and any variety in their excess of energies. In each case, the old motion must be given, and the resulting motion from the composition of the first superinduced force must be found, and this will then become the given motion. This must then be compounded with the motion the second superinduced force would secure as its resultant; and this is then a given motion to be compounded with a third superinduction; and thus onward to any number. The resulting motion must ever be the compound of that which either force applied in succession would give, together with that which had before been given in the original, or any aggregate of superinduced forces. The first principle determines the motion from the perpetuity and constant direction of the excess of energy which generates it; and the second principle determines the motion from the compounding of the aggregate excess of energies in all the forces that conspire to generate it. The law is necessarily given in the eternal principle read by the insight of reason.

The *third* Principle of motion may be expressed in the formula, *that the rate of motion will be as the force moving exceeds the force moved.* This is perhaps more technically given by saying the velocity will be as the dynamical exceeds the statical force. The static force is the intensity of energy with which the antagonism holds itself in position; and the dynamic force is the intensity of energy in one side of the antagonism, by which that antagonism is carried out of its position. In the static, both activities equally energize and resist each other, and the degree of the energies which rest against each other is the measure of the force. In the dynamic, both activities energize and resist, and thus constitute a force; but one activity is of superior energy and thus perpetually displaces this force, and the degree of this excess of energy is the dynamic force. These may be of greater intensity in each point of a small body, so as to equal a less intensity in the many points in a large body; and thus it must follow that it is not the volume but the density of matter that resists motion, and that it is not either the volume or the density of matter, but the excess of energy on one side, that overcomes rest.

In the first principle we had uniformity and direction of motion; in the second we had variation from original uniformity and direction; and here in the third, we seek the degree of motion, or the velocity. Resistance to motion is as the density; and resistance to rest, or capacity to generate motion, is as the excess of energy on one side of the antagonism, and this excess of energy must be most in the densest bodies moved; it is thus mainly with the density of matter that we need here to be conversant, and to find in this the ground for the determining principle of the motion we seek.

The intensity of antagonism in any point of force is its measure to resist motion. If this intensity be small, a small measure of excess in the energy of one activity over the other will generate motion; and if this intensity be great, a greater excess of energy on one side of the activities must be necessary to generate motion. If then one point of force is to move another point of force, the former must have one of two prerogatives; either a greater intensity, and then when just moved its impulse will overcome the intensity of the latter and displace it, or, a strong excess of energy in one side of its activities that may move to a violent impulse, and then, though of less intensity, the strenuous movement of the former may displace the latter. In either case, the principle is at once seen which determines this third case of motion. The force moved is as its static intensity; the force moving is as its static intensity combined with its excess of energy on one side, and however this be made up so as to exceed the force of the former, or force moved, whether by more static intensity or more excess of energy in one activity, when thus exceeding it must generate motion.

And the rate of motion, or velocity, must be proportioned to this excess of dynamic over the static force. The least degree beyond equilibration of intensity must move; and the augmentation of preponderance must so much more move, and thus as nothing but this excess generates motion and all the excess generates its own measure of motion, the degree of motion, or velocity, must be as the moving exceeds the moved intensity of force.

And this is manifestly applicable to all cases. If one body, or aggregate of forces, is to move another, the points of static antagonism are all to be overcome; and the points of static antagonism, in the body that is the mover, all give their intensity and their excess of energy on one side of the static antagonism, that they may conspire to the moving; and thus the aggregate forces are each as one force, and the whole body moving may be called the force moving, and the whole body moved may be called the force moved, and then the third principle of motion is directly expressed by them. The true idea of static and dynamic forces contains the principle which necessarily determines this third case of motion.

In this third principle of motion there is involved the conception of *momentum*, which on account of its wide application in physical science, it is important should be made clear and exact. In the body moving, its power of impulse or capacity to act on other bodies is an aggregate of force from two sources. It has received the excess of intensity over its own in the body moving it, and this now becomes one part of its force to strike and move another body. This is measured by its own velocity, for it is this excess that has made the whole movement, and we may thus represent the force acquired by the *velocity* imparted. But its measure of intensity that it originally had, and which had neutralized just an equal amount of intensity in the body which impinged upon it, has not at all been annihilated. It neutralized its own measure in the other body to produce motion, and left only the excess to pass over into the moved body, but itself remained in, and goes along with, and indeed is the very essence of, the moved body, and this original intensity it now has also, wherewith to strike and move other bodies. This original intensity of antagonism is its *quantity of matter*.

The aggregate of force in the excess imparted from the moving body, and which is represented by the acquired velocity together with its own original intensity of antagonism, and which is its quantity of matter, now constitute the capability the body possesses to generate motion in some third body; and this whole aggregate of motion-generating force is what we comprehend under the term *momentum.* It is commonly said to be compounded of the velocity and quantity of matter, but it should not thereby be understood that mere motion has itself any moving force, or capacity to generate motion, but only that the motion is the index of the moving force which generated it, and which has been transferred to it from the force moving it.

The principle involved in *virtual velocities*, where the less quantity of matter balances the greater, or more generally in all cases of equilibrium, refers at once to the conception of momentum. The less force balances the greater, because the motion of the less would be the more rapid in the inverse ratio of its comparative weight. The momenta of the greater and smaller weight are equal, and though there is now rest, yet is there what is termed *virtual motion*, for the forces are so arranged that if moved, the excess of velocity in one must compensate for the excess in intensity of the other. All static forces have this *virtual* motion, viz., a tendency to move while reciprocally balanced.

These same principles for the above cases of motion, and the conception of momenta and virtual velocities determine all the facts in the pressure of fluids in Hydrostatics, Pneumatics, &c., as also in the revolutions of planetary bodies, and may in the sequel be seen to condition and thereby to give law to all the natural operations of magnetism, electricity, and indeed to be the very nature and essence of the force of gravity itself. The one simple conception of a space-filling force, as an antagonism of simple spiritual activities, is the source in which the reason, without experiment or discursive conclusions in judgment, by its own insight may read the necessary conditions and immutable laws of nature. Force itself is a fact; a thing made; and in its making the very essence of the material world is created; but the immutable and necessary principles which must determine all its working were not made; they lay uncreated and eternal in the bosom of the Absolute Reason, and were the grand archetypes which guided his creative hand in first setting the circuit of the heavens on the face of the primæval abyss.

7. Creation a Nature.—Nature, *natura*, (*à náscor*,) is a birth, an outspringing, a growth; and includes the conception of an existence that has a beginning, and which from the beginning continually grows out, or develops itself, by a successive series of changes which manifest themselves in new phenomena or events. It is a perpetual succession of new births from itself. All these outgrowths were originally in the created existence, and virtually or potentially had their being in the first moment of creation, and necessarily develop themselves in their order as the created existence works on before the inner force of which it is constituted. It is applied properly to every created individual thing, inasmuch as each separate thing has its own peculiarly constituted forces which make it to be what it is, and give to it its own essential identity, and which secure that it must develop itself after the conditions of its original constitution. Hence we say of any particular thing, that it grows, or works, or moves in any way, according to its own nature. And as each constituted or

created thing has its nature, so all creation is, in the same way, spoken of as having a nature. The universal ongoing of cause and effect is but the successive birth and growth of that which was already constituted and necessitated in the first creation, and we thus speak of it as the order of nature. The word has no proper application to that which is not continually passing through an ordered series of births, according to the conditions imposed upon it at its creation. That which was not created, or constituted of such conditioned forces, has not a nature, but must be wholly supernatural. Of all created existence we may say in general, it is Nature.

The propriety and truth of this is seen by the eye of reason, in the Idea so far attained of what creation is. An antagonism of simple activities, which takes and holds position and fills space, and thus constitutes that force which is the essence of matter, combined with a diremptive force that may work in it, has already within itself a nature, and by its creation it already exists under conditions and laws which determine both *that* it must, and *how* it must, produce itself onward in perpetual outgrowths, until its whole inner energy is exhausted. The principles of motion and momenta are principles older than matter, and as unmade themselves may be said to be put into matter when it is made, and thus give to it a nature which wholly conditions it, and which enstamp upon it its whole history, to the penetrating eye of reason, before that nature has gone a step onward in the march of cause and event. Nothing can come out that was not originally put in, and what was originally put in must come out in the very order of the constituted conditions. In nature, non datur *casus*, *i. e.*, events without cause; non datur *fatum*, *i. e.*, events without a conditioned cause; non datur *inertia*, *i. e.*, a cessation of working; non datur *saltus*, *i. e.*, a leaping over some link in the series; and non datur *vacuum*, *i. e.*, a chasm or void within her sphere, where there is utter emptiness.

This does not exclude the repeated interpositions of the Creator. New creations put into nature from a source above nature, and new modifications of the old by the absolute Maker of nature, have within their conception nothing unreasonable. Creation may be finished by any number and distance of intervals between the working, and when finished may receive any number and variety of miraculous interventions from its author, according to his good pleasure. But nature herself can originate nothing, and only bring out that with which she teemed on the very morning of her creation. If the Creator originate new existences in nature, as Absolute Reason he will have reasons for it, and will superinduce the new upon the old in conformity of natures between the new and old, so that the last day's work shall still make one harmonious nature in combination with the works of all other days of creation; and the whole completed work shall as truly fill out and equal the eternal archetype, as if it had sprung up at once and instantaneously under his creating hand. Thus all within and successively coming out below the point where force, as matter, fills space and grows on in time will be nature; a perpetual springing out from the old stock of new modes of existence; but all above that point will be wholly unconditioned by that which is within and below, and will remain forever the supernatural; the unborn, changeless, and absolutely independent I Am. All within nature we may term the *physical*, and all above nature the *spiritual*.

8. The Material Creation a Sphere.—If a force be steadily applied to a heavy body, it will at first be still motionless, but a continued strain at length puts the whole in motion. If I crowd against a boat floating by the wharf, I must perpetuate the pressure for some considerable time before the boat will move. Each point in the body to be moved is a static force, holding itself in its position by its own antagonism, and the force applied must pass from the point of immediate pressure successively through every point to the most remote, and it is only when the last is reached and overcome that the whole mass can be ejected from its place. The force has been constantly going in to the mass, but it has been apparently dormant, or truly latent, until the whole pressure upon the centre of the body has been overcome, and then the mass moves off together.

If I press two rigid metallic rods together at their ends, the force does not continue merely at the point of contact, but propagates itself through every point of both rods to my hands at the opposite ends, and then, though every point in the rods has been pervaded by the applied force, yet do not these points move from their positions; because the middle position at the point of contact in the rods has been equally pressed in the direction of the rods, and the balanced counteraction has kept that point a static in the direction of the force; and also the rigidity of the rods at their points of contact has been greater than the lateral pressure in the compounding of the energies at the middle point, and thus the middle point could not divide itself every way and permit the positions in the lines of the rods to range themselves every way about the middle point as a centre. In other words, the force in the direction of the rods has been balanced, and the adhesion of the metal has not been overcome so as to permit the second law of motion to send off positions in any compounded direction. But if I should procure a complete fusion of the metal in the two rods at the point of contact, and thus dissolve the rigidity, the pressure in the direction of the rods would permit that each should be turned back upon itself by the other, and also, in the compounding of the energies, that each should send off positions every way on each side of the middle point, and the result of the pressure of my hands, in crowding the rods together at their ends, would be an accumulation of the metal from both in a rude globe of molten matter about the point of contact.

A careful application of the principles of motion, or rather the same insight into the principles of force which determine the laws of motion, will detect the very lines in which the molecules of melted matter must move off from the centre, and the very positions they must ultimately assume, and can thus beforehand determine that a globe, and not any other form, must necessarily impress itself upon the matter that shall accumulate about any point of simple counter-agency. For it is to be carefully noted, that not the force which is the component essence of the matter itself in the metallic rods is to be here regarded, but the newly applied force which generates the motion that arranges the particles of matter into a globe. Were there nothing but the forces acting in the matter, whether the rods were rigid or molten, every point would be and remain a static, and the whole mass would be at rest in its original position. The newly applied force is a distinct thing from the force as the matter to which it is applied, and if this were as palpable to the sense as the matter it moves, we should at once notice the force as the mover, and not the metallic forces as the moved. The metallic matter, as palpable to sense, is used by the force applied to it, and which applied force, as working only in the matter, is impalpable to sense, and thus the palpable matter is really only the index of what the impalpable force is doing; the latter registers itself and thereby manifests itself in the former, and it is really only the latter that we regard and wish to follow in its work, for this is truly the sphere-forming agency.

This may be more fully illustrated, and the insight of the reason assisted more clearly to apprehend it by taking two analogous cases, in one of which the matter is apparent to the sense, and in the other the matter does not become apparent. A stone drops into the lake, and as it sinks from the surface it displaces its bulk of water and then passes from that position to a lower, displacing again an equal bulk, and thus successively till it rests upon the bottom. The separating and coming together of the water suddenly through the vacuum left by the descending stone makes a new counteraction, and is truly the introduction of another force into the essential forces which constitute the water itself in the lake. Now this new force registers itself at once in the placid surface of the lake, and we follow its agency in the circling waves that go off from the point of the stone's descent, and in this undulating disturbance we have the visible index of what this newly introduced force is doing. It manifests itself to sense in the effects it produces upon the material forces already existing, and we rest upon the index as appearance, and do not attempt ordinarily to trace by the eye of reason the invisible force which has been really the efficient agent in this circular undulation. But we will now follow the stone in its descent, and as it sank below the surface, there was a like displacement of the water and a coming rush together again through this perpetuated vacuum, and thus truly another central waving expansion impressing itself upon the water beneath the surface, and so on in continual succession to the bottom of the lake. The whole water from the surface to the bottom has been made to arrange itself in circular waves about the path of the stone's descent, but no index of the action of the force below the surface has been given. We see the waves on the surface, but the waves below do not appear. We no longer rest upon the senses, but if we attempt to read the action going on below the surface, we are obliged at once to resort to some new process for apprehending it.

The man least capable of insight, and most dependent upon sense in his activity, will doubtless construct in imagination such circling waves in the water down to the bottom, and apprehend the action of the new force only in the indices which his imagination supplies. It will thus be, as is his apprehension on the surface, solely in the indices of what the applied force is doing, and not in any rational insight that follows the efficiency itself in its working; the only difference being this, that on the surface he sees it, and below the surface he imagines it. But the man that has made himself more independent of sense, and competent to use the insight of his reason in apprehending the efficient force itself, will not care to call in any aid from his fancy in constructing imaginary circular waves that he may follow to the bottom, but he will apprehend that newly induced force itself, and follow it directly out in the waving circles it must make in its own action; seeing immediately in itself what it must do, and not in any constructions of the fancy the indices only of what he imagines it has done. And that mind can follow the naked force in its action upon the surface as well as below it. Independent of what his eye perceives in the produced waves, his reason directly knows the forces and their laws which are there working, and in the forces knows what the circling waves must be, and not merely in the perceived waves judges what the forces have been.

So would it be again, if we put the force at work in the unseen air about us. The percussion of solid bodies, or the force of the human voice, make their similar circular, or, as entirely surrounded, their spherical waves in the atmosphere, and the empirical observer because he cannot see the waves and yet cannot guide his movement without some constructed indices, makes these waves in his fancy, and thus follows the forces where he imagines they have gone, while the mind accustomed to use and rely upon his rational intuitions, dispenses altogether with any constructions of the fancy, and makes the force itself in its own laws pioneer the way to the conditioned appearances which he knows must come from its working.

This following out of the action of force in its own laws is what we now need, and if reference be made to empirical appearances as the registers and indices of the efficient realities, it is only thereby to assist the less independent thinkers, guided awhile by the indices, ultimately to mature the use of their rational apprehension, so that they too may dispense with the sense-perception and guide themselves solely by the clear insight of the reason. In this way we come to know, not by any inference from one appearing fact what according to former experience was the probable fact that preceded it, and thus, at the best, only creep up from one fact to another on the ground of an assumed uniformity in experience, but by an immediate insight into things themselves, we know what those things must be and do, and how the facts must stand from the eternal principles which condition them.

Taking then the independent action of force, as the conception of two countervailing spiritual activities, and following out the action directly according to the necessary laws of motion, we come to the knowledge that matter must accumulate itself about the point of counteragency in the form of a sphere, and must take on all the properties of a solid globe, which has the whole space filled from the centre to the circumference with the successive forces, in their contiguous positions, sent off from the central action of the original simple antagonism. Whether we lean upon the indices in the supposed metallic rods, fused at their pressed points of contact, or take, as an independent object for the reason, the simple agency of a spiritual counteraction, in a force that builds itself up about a position that itself has first fixed itself in, the true condition and law of motion and combination will be the same in both, and the real force which does the forming work will be the same distinct efficiency to be followed in each case. Whether it be pure original force, which takes and fills space and accumulates material existence about its point of counteraction, or whether it be some new force introduced into matter, and moving this already existing matter into new forms, the action and motion and formal combinations in each must be precisely the same, and the practised eye of reason can apprehend and follow the pure original force as readily as that which registers itself in already existing forces.

At the point of counteraction each agency must turn its opposite back upon itself, so that there is not merely a counter-working at one point where the agencies meet, as in the inception of the antagonism, but from the very action of the antagonism, the antagonists have made each the other to react upon itself, and press back upon its own line of action, so that not only now is there counteraction where one simple activity meets the other, but each way in the line of action, each activity has been made to react upon itself, and there is counter-agency each way out of and beyond the point of contact, and thus already has there been an accumulation; a growth, a new-birth of forces from the original point of counter-working. And now, were there but the simple law of action and reaction as opposite and equal, the accumulations of force must be in the right line of the original activities, and each one accumulate, by its retorsion from the energy of the other, new antagonisms in itself successively as from point to point it was made to turn back upon itself. Matter would thus necessarily be generated in right lines. But the second law of motion comes in immediately upon the original counter-working, and so soon as there succeeds a reaction in each simple activity, and thus a force fixing upon a new position out of the original point of contact, there comes at once an extended static each way in this line, and thus an excess of resistance over that of a lateral movement from the point of contact. The reagency therefore cannot move directly back in each in the line of the original antagonism, but must be compounded of the forces acting, and thus moving out every way from the point of original contact, and as it were, lifting itself up every way from this point as a centre; and thus the force accumulates, not only back in the line of the original agencies through their mutual reaction, but also, from the compounding of the movement of such accumulation of resistance in that direction, every way laterally from the point of counteraction. While then the simple reacting force would go out in right lines directly back each way from the point of contact, the compounded forces will rise, as it were in a ring, at the point of contact directly transverse of the original line of action.

But again, so soon as the accumulation should thus begin in this ring at right angles to the original direction, the antagonisms of which the ring is itself composed must turn the component simple activities each back upon itself through all the points of force in the ring, just as at first the one central antagonism turned its simple activities back upon themselves. This pushing each its fellow-activity back upon itself, in every point of force composing the transverse ring, must accumulate two other rings of forces, one on each side of the first or equatorial ring, and which will be, in fact, the turning of the whole ring on each side from itself, and making it to flow in newly engendered streams of forces on both sides backward toward the polar points. The continued activity of the central antagonism, kept by the polar points from going back any further in a right line as an axis, must perpetuate this flowing back on each side of the equator, in new generations of forces, till they meet in their respective polar points, and a proper globe is thus formed by a spherical layer all about the central point. This primitive globe is now self-balanced in all its points; but as the central action goes on, it must again push each way in the axis and generate two other polar points beyond, thereby elongating the axis, and in this elongation there comes as before a static rest in the axial direction, and the central working must rise again in a new transverse ring, and repeat a new flow of forces in their rings from the equator each way to the poles, and augment the globe by another ensphering layer, when all again is balanced, and a new elongation of the axis takes place to repeat the same equatorial rising and flowing back to the poles, and so on indefinitely till the reactions in the accumulating forces of the globe balance the energy of the central working, and the globe ceases to grow. An infinite energy at the centre may generate new layers infinitely. The Almighty may make the globe of the universe as large as he pleases, and when he ceases to augment the central action against the ensphered reactions, the globe will have attained its determined magnitude.

The eternal principles of motion determine the universe of matter to a spherical form, not merely, as Plato assumes, because this is the most reasonable as being the most perfect figure, but the most reasonable inasmuch as the insight of reason determines the space-filling force to such result.

It is also manifest that this sphere so formed will be a

concrete unity, and not a mass of separate and disjoined particles merely aggregated in juxtaposition. The central antagonism turns each simple agency back upon itself by a continuous movement, and as this becomes an extended static, it generates the new compounded lateral movement, that rises as it were in an equatorial ring about the middle point of counteraction transverse of the direction in the simple activities, and then flows back on each side of the equatorial ring to the poles, and balances the whole against the polar points, to begin and go over again the same process perpetually till the universal globe is finished. There is not, thus, a single position within the sphere, except the centre, that has been taken separately and independently; but each position has been taken and held by the new force generated and sent to it in a continuous action from the one preceding, and thus every point of force is held where it is, not merely by its own antagonism but by the conjoint action of every other point of force in the sphere. The movement to each, in the eye of the reason, has been through its preceding conditions, and yet as these conditions could admit of no appreciable interval between them, the whole uprising is to the sense simultaneous. The central point of counteragency is thus at once made a ball, whose radii are the centre and one point or position out of it on every side of it, and the continual working at the centre, continually generates new balls within the old, expanding the old as the new are generated within them, and all the layers thus crowded out by the new central creations are in a continuous connection with the new, and the whole globe is held in one as it were by a perpetuated agency that runs through and connects every position. No portion of the material force is isolate from the rest, but the whole ball is a concrete from the centre through its entire sphere.

It thus follows, that no portion of matter in the forces accumulated in this globe, can be reverted back into the simple agencies from which these forces were generated, and thus become again not force but simple spiritual activity and which would be the same as the annihilation of matter, except by a collapse at the centre. While the central antagonism is constant, all the force that has been generated and sent out from it will press back upon it but cannot escape through it. By no way can the created matter be lost except through a dissolution of the central force, and the instant that this central antagonism should cease and the simple agencies counterworking there should separate, the outlying forces in the globe would have nothing to rest upon, and they must all dissolve, and literally,

> "Like the baseless fabric of a vision,
> Leave not a wreck behind."

All matter, moreover, must thus continually remain not only *so long* as the central force holds, but it must remain *as* it went out from the centre. Except as some modifying action follow out after it from the centre, the force that has already gone off in accumulation about the centre will continue unchanged. The way to superinduce new forces into the material universe, and shape to different results those already acting, will be by new or modified forces at the centre. The whole globe is controlled through the central agency. What has gone off, so far as we have yet followed out the creative action, has been only an accumulation of space-filling forces, and the whole globe is constituted of manifold antagonisms, each and all occupying their own positions, and held in them by the one activity that pervades them all. While thus a concrete, it is also manifest that this globe of forces is a perfect static. Any action that changes the equilibrium in one point finds no counter-action, in the even balance of the whole, till it has gone through and equalized itself in the change of the whole. An introduction of any new force, or augmented action, at the centre, must make itself felt through all the globe.

9. The Principle of Gravity.—It is the crowning glory of Induction, that it prompted to the search and guided to the attainment of the law of gravity. The name of Newton is made immortal from this sublime discovery. It will detract nothing from the true honor of the Inductive philosophy, nor from the undying fame of Newton, to put in the precise light of truth what is the exact amount of physical science secured in that discovery. Its hypothesis was that there was a tendency in all matter to approach to all other matter in certain ratios. This hypothesis was suggested to the fertile mind of Newton by a single occurrence, and when tried by extended observation, it was found in accordance with so many other occurrences, that there was no hesitation in assuming it as a universal fact. Very extended and profound researches into appropriate facts, especially the complicated facts of the variations in the moon's revolutions, at length confirmed the conclusion, that the ratio of this gravitating tendency in matter was directly as the quantity of the matter, and inversely as the square of the distance. This fact of the gravitating tendency of matter, and the further

facts of its ratios, put into a general formula, enables the natural philosopher to classify an immense amount of particular facts under this simple category, and ever after to know their place and recognize their relations in the grand system of facts which he brings together. This is the extent of its use, the taking of this as a broader fact which may embrace a vast number of particular facts within it, and enable to say, because this is fact, therefore these other particular facts are as they have been found to be. It is thus called the *Law* of Gravity, not because the principle determining it has at all been found, but only because such a general fact having been found, the subordinate facts can all be referred to it. It does not reveal why the facts embraced in this formula are, nor that they might not have been other and opposite; but simply having by a broad induction so far found them thus to be, and then assuming that through all experience they will be found so to be, there is hence the warrant for concluding that these particular facts are embraced in the universal one. It can only suppose matter, as originally inert, to have such a tendency arbitrarily imposed upon it, and not at all that there are principles older than the facts, and which eternally and immutably condition the facts, and are thus infallible reasons for the facts.

Of this tendency in matter to approach other matter, no explanation can be given. It may sometimes be said that it *seeks* to approach, as if the explanation would be given that there was some sentient life in matter, and that this tendency was the congeniality of social affinities; the movement of matter expounded by the susceptibilities of mind. Just as it was early said in explanation of the rising water in the pump, that nature *abhorred* a vacuum. This *attraction*, as now named, is as wholly vacant of all reason as was the *suction*, as then called. When the fact of gravity had been discovered, then this power of suction came readily within it; the weight or gravitating force of the atmosphere pressing upon the water out of the pump, forced it up into the vacuum made within the pump, and we put away all our conceptions of "the powers of suction" and "abhorrence of a vacuum," and by an insight of reason follow the force which does the whole work, and smile at the unreasoning simplicity of an earlier philosophy. But what has been gained, except simply removing the mystery and our ignorance one step further back? Why the atmosphere seeks the earth is just as truly without a reason, as why nature abhors a vacuum; and the word *attraction* has within it as gross a solecism, when the necessary law of forces is apprehended, as had the old word *suction*. Matter no more *draws* matter, than the pump *sucked* water. The pump removed the air from a space, and an outside force pressed water into it; and so the central force sends off matter through a given sphere, and the same force which sends off all molecules thereby presses back each one. And as an estimate of the force of atmospheric pressure enabled the philosopher to determine the power of the pump, even so does the estimate of the central force enable the philosopher to determine the power of gravity.

The difference in the two cases is, however, very wide and important. The philosophy of the pump was grounded in a higher *fact* only, and that fact left wholly inexplicable. The pump was explained only by a reference to a higher fact which could not be explained; and by leaving the higher inexplicable, the whole was truly an impenetrable mystery. But if we expound the force of gravity, not by running it into some higher fact which is itself left in darkness, but by applying an unmade and eternal *principle* to it, which must necessarily so condition all facts of material existence, then have we a radical and ultimate exposition, and have traced all the mystery in the nature of matter up to the light of the supernatural reason, and can say gravity is thus, because the immutable principle in the absolute reason determined it must be thus and not otherwise. Gravity is then no longer a mere *fact*, a thing made, and which might have been any otherwise made; but a fact with an eternal and immutable law in it; a fact embodying an uncreated and necessary principle of the reason, and in following which principle in the making, the absolute Creator manifested his immutable wisdom and truth. To this eternally necessary and immutable law of gravity, we now turn the insight of the reason, that we may clearly apprehend the unmade principle which conditions and determines it to be thus and no otherwise.

No one point in the sphere can be equally balanced in its own simple antagonistic agencies, except the central point. Here the originating, simple activities begin, and hold each other in balanced energy, and turn each other back upon themselves, thus making a tendency to accumulate force in two positions on the line of direction each side and out of the centre, and which tendency, as we have above seen, creates further the tendency to accumulate in an equatorial ring, and turn back each way this ring till its accumulations make an ensphering layer on each side to the poles. As these new hemispheres of layers on each side of the equator are formed, they balance the whole equatorial antagonism in the aggregate in the polar points, and thus it must be that the edges, so to speak, of the hemispherical layers in the equator push back on each side in meridional lines against the poles. The whole globe is a unit, and yet the two simple activities so push each other back upon themselves, that one hemisphere is generated from the retorsions of one activity, and the other hemisphere from the opposite activity. As the generation of the globe proceeds, each layer, and each meridional line in the layer, must turn itself back from the energy of its antagonist, in new hemispherical layers, and thus the globe grows on the inside in every layer perpetually as well as by an outside layer; or, in other words, the whole globe augments itself by the antagonism of its two hemispheres, as the first central point does by the antagonism of its simple activities.

We may then carry on the intuitive process, and take any position out of the centre and contiguous to the centre, and it will be a point in a layer of points ensphering the centre, and it with every other point of that ensphering layer is pushed out and held in position by the antagonist force working at the centre. At the same time, also, that the central antagonism is pushing out and holding in position all the points in this contiguous ensphering layer, each one of these points is reacting and pushing back upon the centre, and the aggregate of force in all these points just equilibrates the force of the central antagonism. No one force out of the centre balances the centre, but the aggregate of all the forces in the contiguous outlying layer.

So also, take any contiguous point out of this first enspher-ing layer from the centre, and it will be one of many points in another layer enspheering the first, and all the points of this second enspheering layer will react upon the inner layer, as the points in it react and balance themselves upon the centre, and thus the aggregate of force in the second layer equilibrates the aggregate of force in the first layer, and this equilibrates the antagonism at the centre, and thus on through all the concentric layers that may have been pushed out in completing the whole globe.

The central point expels the outlying points on all sides, but each point in the contiguous layer of points about the centre, while in the same way acting outwards on all sides, must on the side towards the centre act upon it, and only on the side from the centre can act upon the layer exterior to it, and in concert with all the fellows of its layer push out this layer beyond. Out of the centre, therefore, each point in every layer acts on one side towards the centre and balances itself upon it, or upon it and the points intervening, and on the other side acts on the contiguous point of the exterior layer, and pushes that from the centre. The quantity of force in every molecule, and the direction of its working through the universal globe, how large soever it may be, is thus determined in the necessary conditions of the working of that central antagonism which generates the universal sphere. So much only we need now to note, that every molecule of force works outward from its own inner antagonism, and by the working of all, every molecule of the universe out of the centre is pushed from the centre, and also is repelled back toward the centre, and the aggregate of all the forces pushing outward is just balanced by the aggregate of all pushing inward, and thus every molecule comes to rest in a static equilibrium.

It is a necessary determination that a globe so generated, should have in every molecular force a centrifugal and a centripetal tendency just balancing each other, and thus holding the molecule at rest. The first is properly *expulsion*, as it is a primitive driving from the centre, and the latter is properly *repulsion* in its return back toward the centre. An utter misconception, as if at first there was somehow a suction drawing to the centre, and then a rejection of the same from the centre, has appropriated the term *attraction* for what is truly the reagency, and *repulsion* for what is truly the first outgoing agency. But any attempted change of terms would now be hopeless, and we shall therefore use attraction for the centripetal and repulsion for the centrifugal tendency, yet philosophically noting perpetually, as in the Copernican system it is not the sun that rises and sets, so here it is not the coming to the centre that is primitive, nor is this coming a *drawing* but a real *a tergo pushing.*

We will then carry the insight of reason directly on through this idea, that we may determine the ratios of these forces.

The central point of force pushes out all points and just equilibrates their aggregate reactions, and is thus truly the measure of force in the whole sphere. Every other point also pushes out in like manner in every direction from itself, and if left to its own action would ensphere the contiguous outlying points of force about itself, but no point of force except the centre can push out and equilibrate all the other points of force in the sphere, since, on one side, all points of force push back upon the centre and balance themselves and all pushing back on them in the centre. As is the energy of the central antagonism, such is its degree of force and, which is the same thing, its quantity of matter; and in proportion to its energy will be the magnitude of the sphere generated, and thereby the degrees of force, or, as the same thing, the quantity of matter in each point of force through the whole sphere. The amount of centrifugal force in the whole sphere is, therefore, directly as the quantity of matter, and the amount of repulsion in the central molecule will also be as its quantity of matter; and as each point out of the centre must be pressed out and press back proportioned to the layers within and beyond it, so all molecules of the sphere must also have their repulsions directly as is to each one its quantity of matter. In all respects, the force of repulsion must be *directly as the quantity of matter.*

The molecule at the centre repels all the outlying molecules on all sides from the centre, with a force directly proportioned to the amount of matter in the sphere, and in the case of a sphere standing alone in the void, as must the universal sphere, the amount of matter in the sphere must be as its volume, or, which is the same ratio, as the cube of its radius. But any concentric layer of molecular forces is diminished in its repulsion in proportion to the number of layers that are between it and the centre, *i. e.*, in proportion to the cube of the radius of its own sphere, and thus each ensphered layer of molecules repels inversely as the cube of the radius of its own sphere; and as each molecule in the layer may be understood as the terminus of a radius to the ensphered layer in which it is, so any molecule in a spherical layer, and thus any molecule in the whole globe, repels *inversely as the cube of its distance from the centre.* We have, therefore, the necessary law for repulsion—*directly as quantity of matter, and inversely as the cube of the distance.*

The force of attraction does not, like the force of repulsion, act from the centre outwards on all sides, but is the reacting force against repulsion from every point in the sphere and coming back in a direct line to the centre. The aggregate of attraction in the entire sphere is equal to the amount of repulsion going out from the centre, for they equilibrate each other in the static position of every molecule. The repulsion is, however, from the point on all sides, but the attraction is from the point on one side only and working directly towards the centre. We may, then, take all the points in any plane passing through the centre, and we shall have the aggregate attraction of that plane in right lines to the centre. This will give the aggregate attraction to be as the area, or, as the same thing, to be as the square of the radius. But any concentric circle in this area is diminished in attraction in proportion to the number of concentric circles that may be made between it and the centre, *i. e.,* in proportion to the square of the radius of itself, and therefore each possible circle in this area is attracted inversely as the square of its radius; and as each molecule in the circumference of the concentric circle may be understood as the terminus of a radius of that circle, so any molecule in any concentric circle of the area, and therefore any molecule in the whole area, is attracted to the centre inversely as the square of its distance from the centre. But this diminishing attraction from the centre vanishes wholly away in the circumference, and thus the attraction, between any two points separately and respectively in the same radius, is *inversely as the square of the distance between them.*

This must be the law of attraction for all points in the radii of the same sphere, for they may all be taken in their relative planes and thereby subjected to the same principle. And this applies not merely to the points in the same globe, but relatively to all globes; for all globes must have their repulsive and attractive forces balanced, and the intensity of these antagonisms constitutes their quantity of matter. As is the pressure at the centre such must be the magnitude of the ensphered forces, and thus such the length of radii, and the distance to which the force goes out from the centre, and the reacting attraction comes back. In all globes, therefore, the attractive force must be *directly as the quantity of matter, and inversely as the square of the distance.*

But this is true, again, not only of all globes in respect to each one's own portions of matter among themselves, but of all globes relatively to each other. Each globe must have its own density and thus its own distance for its force of gravity to act, and when any two globes come within each other's range of attraction so that the peripheries of their spheres cut each other, the point of contact is at once a point of antagonism, and their acting central forces must so work this commencing antagonism as to push each one back upon itself and begin an ensphering anew, with the central point at the first point of contact, and the forces of each globe must be successively turned back in a hemisphere within itself, and both together must form a new globe around this central point, and like "kindred drops both ultimately mingle into one." Such common point will become the common centre of attraction for each globe, and if the matter be fluid the two will make one globe, and if rigid, that point will still hold the two globes in unity and become their common centre of attraction, and must act under the above eternal principle. Any masses of matter, less or more, must stand to each other as such two globes when they have their gravitating forces brought in contact, and their common centre of gravity must work after this eternal principle.

10. The Principle of Falling Bodies.—The principle of gravity being attained, we may consider the force of attraction, though made up of the compounding of all the reactions in the globe, as if it were in the case of each molecule a separate and distinct force. Each molecule tends to the centre as if it had one simple activity of greater energy working toward the centre, and one activity of less energy working from the centre, and as if the side of the weaker energy was helped in sustaining the position of the force against the greater energy, by all the forces in the line between it and the centre. Take, therefore, any molecule of force in any part of the universal globe, except the central one, away from its position, and view it as if standing alone, and it could not be a static; it could not hold its own place; it would be impossible that it should lie still. One side has the greater energy, and the molecule of force must move before that energy, since it cannot rest against the weaker energy in static position,

until it can find some competent assistance to the weaker energy, and thus balance itself at rest.

This idea of a force of unequal antagonism determines the generation of motion, in the advance of the greater energy through successive points. In this is, also, by the first principle of motion, the determination of it as uniform and rectilineal. This uniformity of velocity is on the condition that the excess of energy in the moving activity be perpetual and invariable. But the action of gravity, by which any aggregate of forces tend to the centre, though working in a right line cannot continue of uniform energy. There must be a constant accumulation, inasmuch as there is a constant transfer of new force with a perpetual retention of what has already been received. A full conception of the force of gravity gives occasion for determining this increment of motive-energy, and attaining the law by which the velocity of any body falling unhindered towards the centre must be regulated.

We may assume any point of force, or which will be the same thing, a body with its aggregate points of force, and if the forces in the points of the radius between this body and the great centre be supposed to be weakened in intensity, or those in the body augmented, that body must move through the line of the radius towards the centre, and which is but saying that it must fall. This necessity for falling is seen in the excess of energy which every point in the sphere has, in that activity which is further from the centre and is working towards the centre. When the intervening pressure from the centre, which assisted and thus made equal to the other that activity which was nearest the centre, has been taken away, this excess of energy must prevail and generate a movement in a right line before it. The one activity is not of sufficient energy that the other should rest itself against it, and both in their common point of antagonism must fall together. But such fall cannot be uniform, as in the steady force which determines the first law of motion. There, the excess of energy, or moving force remains constant, and just avails to move at its own rate of velocity without accumulation. Here, it moves and adds itself perpetually. The excess of energy which generated the movement remains, and when at the next position nearer the centre, there is also the excess of energy that would have generated an original motion at that point, and thus the one degree of excess is retained and an additional degree received in the movement from this second position, and in this necessity of perpetual increment is determined the law by which it must fall.

Take then any body, and let it possess its degree of excess of energy on the side opposite the centre, and it must gravitate toward the centre. It must thus pass through its one measure of space, or height, which we will call H, in one moment. In passing through H, it has gained the excess of energy, or gravity, that was in the force occupying that space, and must therefore have now an excess of energy, or gravity, from the fall through H, that would make it fall the next moment through 2 H. And the original excess of energy, or gravity, with which it started is still retained, and will make it pass through one measure, or H, for this next moment as it did the first, so that the body for the second moment must pass through 3 H. At the end of the second moment, there are the three degrees of excess of energy, or gravity, retained, and the last one is doubled in the gain of its gravity, and the original excess with which it started at first is also still there, so that the body must fall the next moment through 5 H. These five degrees and the last one doubled, and the old excess with which the body began to fall still constantly acting, must make the fall to be for the fourth moment through 7 H. These seven, and the last doubled, and the old excess, must again make the fifth moment to have a fall of 9 H. Thus onwards perpetually with each successive moment. The succeeding moment must have all the gravity of the preceding passing over in to it, and the last degree gained in the preceding moment must be doubled from the excess of energy gained and retained, and the old primitive excess is perpetually going along, so that each succeeding moment must continually gain two measures of space, or 2 H, above the preceding moment. This must be perpetual, so long as the body continues to fall. The principle is in the force of gravity itself, that must make an increment of two measures of descent, or as the same thing, two degrees of velocity, in each successive moment. The first moment will have one degree of velocity, the second moment three, the third moment five, the fourth seven, and thus on in arithmetical progression perpetually with an increment of two degrees to the moment.

There must thus be at the beginning of each successive moment, an excess of energy, that would carry the body through double the measures of height fallen in all the moments for the next succeeding equal number of moments. Thus at the end of the second moment the body has fallen through 4 H, and for the next two moments would fall through 8 H, and at the end of the third moment it has fallen through 9 H, and in three more moments would fall through 18 H, and at the end of the fourth moment it has fallen through 16 H, and in four moments more would fall through 32 H, and thus onwards perpetually. This will secure that, at each moment, the whole measures of space fallen, or the degrees of velocity attained, must be directly as the squares of the times, and which, in the opposite direction from the centre, must also be demanded by the law of gravity to be inversely as the square of the measure of space, or distance, or in this case the velocity. Thus the first moment has fallen 1 H, which is the square of itself; the second moment 4 H, or the square of two; the third moment 9 H, or the square of three; the fourth moment 16 H, or the square of four; and so onwards indefinitely. Thus the sum of the laws of falling bodies as determined by the principle of gravity is —that each moment must increase by two degrees of velocity—that at the end of each moment, a velocity is attained that must fall through double the space in the next equal number of moments—and that the velocity gained is directly as the squares of the times.

The same principle, modified by the cutting off and neutralizing a part of the accumulating force, is found in the descent of bodies down an *inclined plane*, and thus determines the law for the increment of momentum, and the counteracting force necessary to raise or balance weights upon an inclined plane. A body placed upon an inclined plane is acted upon by the same force for generating motion, as in the above case of a body falling freely. The excess of the antagonism tends to motion in the perpendicular direction towards the centre of gravity. But the interposition of the plane, more or less obliquely, cuts off and neutralizes this force in its direct action, and proportioned to the angle of inclination dissolves and diverts a portion of the force in the direction of the plane. If it had moved freely from the start in the perpendicular direction, it would have been wholly within the former category of falling bodies, but now a proportion of the moving force is neutralized by the degree of inclination given to the interposed plane, and only the remnant in the turned direction is in action to give and gain an increment of momentum. The *perpendicular* descent would begin with the whole energy of the excess of static force, and observe the laws of increment of momentum in its progress; the descent in the *inclined plane* begins with the remnant that is not neutralized by the interposition of the plane, and observes the same law of increment of momentum in falling down the plane. The plane is the hypothenuse of a right angled triangle, of which one of the sides containing the right angle is the perpendicular, and the other containing side is the base of the plane. The times and proportions of increment of momentum are, thus, as the proportion of the perpendicular to the hypothenuse. The descent must be through the whole plane to gain the increment of relative momentum, which the body falling freely gains in descending through the perpendicular.

It thus follows that a proportionally smaller body, falling freely, may be made to balance a larger body falling down an inclined plane. Two bodies, so attached that the smaller shall act through a line in the plane in opposition to the descending force, and left to fall freely down the perpendicular, must equilibrate when equal momenta are gained in equal times. The disparity in the quantity of matter of the bodies balancing, will be as the difference in the rate of movement; the less velocity on the plane being the index of the lesser increment of momentum in the same space, and the greater velocity down the perpendicular being the index of the greater increment of momentum in the same time. Here is no gain of force, but the economy of substituting a less force for a greater period of working; but this economy is, however, of great moment, and may be used to an indefinite extent. The universe of matter may readily be conceived as thus balanced by a grain. The principle of the inclined plane is in the wedge, which may be made to incline on one or on both sides, and in the screw, as a spirally inclined plane, and thus all these so called *mechanical powers* find their full explication in the conception of momentum as attained solely through the insight of the reason in the law of falling bodies, connected with the conceptions of momentum and virtual velocities.

Applying the same representative of the increment of velocity for the increment of force which generates this velocity, in what is termed *virtual* velocity, we may in the like manner attain the principle of the lever, as a mechanical power, in all the ways of its application. If we conceive the diameter of a circle to be an inflexible rod resting upon its centre, and that masses of matter of equal quantity are affixed to the two extremities, it is plain that they must balance each other, inasmuch as their momenta, *i. e.*, their quantities of matter and virtual velocities, are equal. But if we shorten one semi-diameter, or slide the rod on the centre to make its two ends unequally distant from it, the virtual velocity in the longer, it is at once clear, must be the greater. To preserve the equality of momenta, there must be a corresponding augmentation of the quantity of matter on the shorter, or diminution on the longer arm of the rod. This supposed rod is the lever, and it is evident that the same principle is involved here as in the inclined plane, with a different phase only in its presentation. The inclined plane brought the greater weight to the same height by a longer *way*, and thus gave opportunity for the accumulation of the excess of force in the smaller weight; and the lever brings the greater weight to the same height through a longer *period*, and this gives also opportunity for the accumulation of the excess of force, and thus an increase of virtual velocity to the smaller weight.

The principle of the lever is in the wheel and axle, and also in the pulley, and these together may be indefinitely compounded in cogs and bands and tackle-blocks, and all together may be combined with the varieties of the inclined plane, and thus give endless conveniences, but in no case any creation and only a transferred accumulation of power.

11. THE PRINCIPLE OF MAGNETISM.—The central force from which the universal globe of matter is generated, it has been seen, necessarily induces a tendency to motion in two opposite directions equally balancing each other, viz., propulsion from the centre and a reaction in each point of force which presses back in every way towards the centre. A careful insight into the working of this central force will also detect another virtual movement in its necessary principles, and which, when fully apprehended, will be recog-

nized as filling out a full idea of the force of magnetism. A globe, so formed, must be a magnet.

If two balls of melted metal were brought in contact, and gently pressed so as gradually to flatten each up to their centres, a peculiar process would necessarily be passed through in this changing of forms. When just brought in contact, and touching at but one point in their circumferences, the antagonism would be as that of two simple agencies. One ball would not at all press itself into the other, but each would turn back the opposite ball into its own body. As the pressure went on, each ball would have a retorsion of itself within, of just the same size and shape as that which had been displaced from the flattened segment, and when the centres should meet there would be a ball within of the size of the original balls, of which one hemisphere would be from the retorted portion of one of the original balls, and the other hemisphere from the portion turned back in the other. If the original balls had been of two different metals of equal densities, or of two different colors, one hemisphere of the new ball would be of one metal, or color, and the other hemisphere of the other.

To make this the more manifest, where we may follow the forces in their sensible indices, we drop again a stone into a lake. As the circle of undulations expands, at a little distance from the centre we will conceive an obstacle interposing itself, as a rock standing upright from the bottom. The periphery of the circular undulation, just as it touches this obstacle, is the index of the simple agency in one direction, as now acting in a right line from the point where the stone descended, and the resisting rock may represent the other simple agency at the point of counteraction. When the forces thus meet, the action and reaction being opposite and equal, the tendency at first would be to turn the force in a direct line back upon itself towards the point from whence it came where the stone fell, and give its index in a refluent wave in this right line. But as the re-agency goes back it gathers additional resistance in the attempted refluent wave, and which would perpetually increase in each moment of the regressus. It thus at once supervenes that the action and reaction cannot be opposite and equal, in a point removed back from the rock towards the place of the fallen stone, but the resistance to go back must be greater away from the rock than at the rock. The law for the compounding of forces and motions at once controls, and the result is, not a refluent action in a direct line, but an action spreading each way from the rock on the side of the counter-agency, over the surface of the lake. The first stroke of the outgoing wave in the circumference of the circle upon the rock makes its refluent wave, and the next stroke also its refluent, expanding as it flows back with just the force in the aggregate that had brought the circle up, and thus the result is, a return of just so much of the circle as would unimpeded have gone on by the rock now within itself on the hither side of the rock, and this action may be perpetuated until the refluent has equalled and exhausted all the up-coming waves. What has gone back is that which would have gone on, and it has now registered itself in the matter on one side of the rock instead of the matter that lies on the other side, and thus the real thing, as force, has moved and fixed itself in a retorted position. If the rock could have been the force of another circle coming up in antagonism, and they could have been made directly to meet and counterwork without slipping past in the matter that brought them together, there would, as in the melted metal above, have been the formation of two inner semi-circles, or if the action had been sufficiently deep in the lake, two inner half-spheres, and each having only the force in its own hemisphere that had been pushed back by the force that has gone into the other. Thus two circles, or spheres, make a third, with one-half within and of the one, and the other half within and of the other.

And here, it is to be noted, precisely the same thing must occur in the generation of forces into a sphere about the point of two countervailing single activities. One turns the other back upon itself, and the accumulating resistance in a right line necessitates a compounded movement and fixing of positions every way in a sphere about the point first brought in contact, and one-half of this globe is generated in the forces on one side, made by the return upon itself of one simple activity, and the other half by a return upon itself of the other simple activity.

And here it is practicable to the eye of the reason to follow these moving activities, and see just how they must result in fixing the matter which they generate, or in leaving their register as an independent accumulation of forces, in any preceding matter that might be given, as a medium for indicating their currents. If we take the case of two circles in a lake, there might be conceived two floating rods lying as a tangent to each circle, and in the same straight line with each other on the peripheries of the same sides from the centre. As we view these circles and the rods pointing towards each other, we might say of the one on the right hand that its point was the boreal, and of the one on the left hand that its point was the austral, in reference each to those points in the undulating circumference of their own circles as about to touch each other. If we now conceive these expanding circles to meet, and make each the waves of the other to be refluent within itself, we shall have a new circle made up of two semicircles from each, and within each, about this point of contact. If we should conceive of the augmentation of this new inner circle, by the approach of the original circles together, until their centres should meet in this point of their first contact, and which has become the centre of the new inner circle, then the floating rods would be brought together at the circumference where the half of the original circles now meet and cut each other, and the boreal point in one, and the austral point in the other, would each be retorted and turn each the opposite way in the refluent semi-circles of the new circle formed within and from the original two circles. At just the point in the diameter where the two semi-circles now make one circle, the two rods would be as one, and having both a boreal and an austral point, and lying as a tangent to the circle at right angles to the diameter. Thus with a circle on the surface of a lake; and the same result would occur in the sphere, which might be formed by two refluent spheres down within the lake. And this analogy is also perfect in reference to the space-filling matter, which first enspheres itself about an original point of counteraction. *The hemispheres must have a bi-polar force at the equator.*

This index rod, which now may lie on any part of the equator, or, as the same thing, at right angles to any equatorial diameter, will lie parallel to the axis of the sphere, and may be made to point indifferently either way, inasmuch as it is really the identification of the two original rods, and may be either a boreal or an austral point towards the same pole. The force, which registers itself in this index, is a retorsion on each side of the equatorial ring towards the poles, and thus directly on this ring is the mutual and neutralizing limit between the reactions, which must be indifferent to either. But as there is a departure from this mutual limit either way, the refluent force must take it on its own side, and control it by its own movement.

How it must determine its direction is plain by careful inspection. At the equator, the forces are on each side at right angles to the diameter and parallel to the axis of the globe, while at the poles, the forces are in a point reactionary to the central antagonism, and thus up and down within the axis. The directions of the force on the side of the equator is in a parallel line with the axis, and the direction of the force at the pole is towards the centre in the line of the axis. The direction of all the forces on the side of the equator are thus to meet in the common polar point, and become turned in that point to a line in the axis towards the centre, and there must in this be compounded the forces which work from the equator, and the conjoint force in the polar point which works towards the centre. This compounded movement, it is plain, will be a perpetual turning of the index from a tangent to the circumference at the equator, to a direction that must go athwart the circumference, and thus a perpetually increasing dip must be secured, making it at right angles to the axis at the distance of 45°, and in the direction of the axis, towards the centre when over the pole. This will be the same with the austral dip on the boreal side of the globe, and with the boreal dip on the austral side. The forces of the sphere necessarily determine for it *a magnetic polar direction and dip on each side of the equator.*

We may also look immediately into the operation of the sphere-forming process as given in section 8th, and we shall attain a similar, and even more comprehensive result.

The central antagonism turns each its opposite simple activity back upon itself, and thereby generates the polar points; these hold the energizing of the activities from pushing further back in a straight line, and thus the equatorial ring is at once elevated; and then the perpetuation of the central energizing pushes the simple activities in each point of the equatorial ring each way, in meridional lines, quite up to the poles, and all such meridional lines in their contiguity make a spherical layer of molecular forces over each hemisphere, and in this continuation of working, the globe grows to its determined size. We will take these points of molecular forces in the equatorial ring, and subject their determined progress to a careful insight.

Each point of force in the equatorial ring is an antagonism of two simple activities counter-working each other in a direction parallel to the axis, and thus the whole equatorial plane is made up of antagonist forces, all working in their directions of counter-agency parallel with the central antagonism, for the aggregate of points in the whole equatorial plane are but the edges of the concentric layers as they meet in their equators. The points nearest the centre will be in the equatorial edge of the smaller, and those further from the centre will be in the equatorial edges of the larger spherical layers. If, then, we take any point in the equatorial plane, it will be a point of force in the equatorial edge of some spherical layer, and will balance itself back upon the polar point in the direction of the meridional line, and by the equilibration of forces that lie in and all through that line. We take, then, any such point, and consider it as a molecule of matter that has its polar direction in the line of its two counter working simple activities, and which must be parallel with the antagonism of the central point, and with the axis of the universal sphere. The central point works back upon the polar point in an exact line of action and reaction through all the intervening points, and thus each molecule is in the same polar direction as all the others in the axis on its side of the centre, and all on one side of the centre of opposite polar directions to all the molecules on the other side of the centre. But, while the central point thus works exactly back in molecules of the same polar directions, no point out of the centre can work back upon the pole in molecules of the same exact polar directions. The point out of the centre, and in the equatorial plane, begins with a molecule that has its polar direction parallel with the centre, but the polar point against which it balances has its polar direction towards the centre and in the line of the axis; this beginning equatorial point must thus have the direction turned gradually and completely about, in the molecules of the meridional line, by the time the action has reached the polar point and balanced all the molecules in the meridional line upon it. The equatorial molecule will have its polar direction parallel with the axis; the polar molecule will have its polar direction in the axis towards the centre; and thus all molecules in the meridional line between the equatorial and the polar molecules must have their polar directions conformed to their respective positions in this meridional line. The next from the equatorial must converge from a parallel to a slight inclination of polar direction towards the axis, and the next to that a little more, till midway, or at the 45th degree, the polar direction of the molecule must be at right angles with the axis, and so onward turning about to the pole, where the molecule resting on the polar point must have a polar direction turned completely round and working towards the centre in the line of the axis.

This principle of polar direction in the meridional molecules must necessarily determine the magnetic dip, and this must begin from a parallel with the axis in the equatorial plane throughout, and terminate through the meridional line by a direction to the centre in the axis at the pole, having in that distance completely retrograded its former direction. The opposite poles must have their magnetic dip the converse of each other through all the molecules of their respective hemispheres.

Bring, then, two magnets together, and force their spheres of polar action to invade and interpenetrate each other, the similar poles of each would have their molecular action and dip opposite to each other, and each towards its own centre, and they must repel each other; while the opposite poles so interfering would have the molecular action and dip conformed to each other, and they must therefore attract one the other. The molecular action and dip is neutralized in the equator, and thus mutual attraction and repulsion must there be neutralized, and the respective attractions and repulsions must augment to their maximum in the interfering poles.

Thus the equatorial and polar counteractions determine the necessary laws of magnetism, as the central action and reaction determine the necessary laws of gravity.

12. THE PRINCIPLE OF ELECTRICITY.—The balanced action, between the equatorial plane and the poles, which holds every molecule in its place of static rest through every meridional line of every spherical layer, is the great principle of magnetism; the interruption of this continued static rest in any portion of the superficial matter, and the consequent tension in the interrupted parts to recover themselves and restore the balance, is the great principle of electricity. This pressure from the equator to the pole, and in the stability of the polar point a reciprocal pressure back from the pole upon the equator, we have now to conceive may be interrupted in particular places by various agencies. These several interrupting agencies will be hereafter examined in their determining principles, such as light and heat and chemical decomposition, and to which may be added mechanical friction, the principle of which has been already apprehended; but these causes and their occasions of working in the interruption of the magnetic force need not here be regarded. We need only anticipate, that causes and occasions will occur to break up this continuity of equal reciprocal agency between the equatorial plane and the polar points, and that thus the molecules will in some places be turned in their polar directions away from their proper lines in the magnetic meridians, and in this

anticipation we shall find all that is necessary for clearly attaining the grand principles which must determine all the laws in the experimental facts of electricity.

Some substances, then, in the ongoing of the forces of nature, must be supposed to be brought into combination, that will admit of their respective component molecules being more or less easily turned from their proper magnetic polar direction, as would be their natural position in the magnetic meridian, and when thus deranged, some of these substances will be very slow and stubborn in permitting their molecules to come again into their proper magnetic arrangement. Let such deranged substances occupy any place in the sphere between the equator and one of the poles, and they must at once interrupt the reciprocal action between the equator and the pole and sunder the magnetic meridional continuity. There will then necessarily at once ensue a tension in the equatorial force to overcome and remove this interruption, and to rest itself again, in the continued line of magnetic reciprocity, upon the resisting static force that comes up towards it from the pole. Such struggle and tension to overcome this interruption is the awakened and active force of electricity. It is really the magnetic force struggling against interposing and interrupting derangement. Magnetism and electricity differ, as static polar rest differs from the tension that struggles to remove an interruption that it may again be at rest.

Now such tension must manifestly have its two conspiring directions; the equatorial force will go out positively and actively to find its reciprocal static point and rest itself against it, and this static polar point, unsustained by its positive antagonism on its equatorial side, can only negatively struggle for such sustaining reciprocity, by maintaining itself against the central antagonism that would go beyond it in fixing a new polar point, and force this central activity to go up in the equator and come down in the meridional line to the pole, and there meet and support itself in counteraction. There must, therefore, be two forces; one positively struggling to move forward, the other negatively struggling not to move back. We shall therefore have two kinds of electricity, properly the *positive* and the *negative;* and no interruption to the polar force can anywhere occur, but it must at once induce this positive and negative tension. The deranged substance, interrupting the magnetic continuity, has its molecular polarity all the wrong way, and each molecule turned so that the end of its axis which should be towards the pole is now toward the equator, and thus this whole interrupting substance is turned with its positive towards the positive tension that struggles against it, and its negative towards the polar negative that holds itself not to go back, and therefore, the positive tension must push or repel opposing positive tension, and negative tension must stand against opposite negative tension; and, on the other hand, the positive must constantly push into the negative that is before it, and the negative constantly receive the positive that comes up to it; and thus the fact in experience must ever be that opposite electricities will attract and similar electricities repel each other. This is similar in expression but not in principle to the attraction and repulsion of magnetism. In magnetism the whole magnet is a unit, and the equator or middle point neutral, and the polar forces and dips are opposite to each other on opposite sides of the equatorial plane, and thus here the principle is, that similar dips and forces must throw out, and opposite dips and forces must crowd in each other; but in electricity the whole has no unity, for this has been broken up and the entire force of electricity is to recover it, and the opposite electricities work in the same directions, one positively and the other negatively, to join and restore the unity. There is no centre to the electric tension; all is in one direction, and hence any division of an electric cannot make the parts to be now complete wholes, as in a divided magnet, but must still be each one only another fragment of the whole.

It is also further manifest, if any substance may be made to penetrate this deranged molecular substance that interrupts the magnetic continuity, and such penetrating substance have all its molecules facile and ready to come to their proper polarity, that this penetrating substance, uniting the positive and negative electricities through the interrupting molecules, will at once bridge over the impassable chasm, and in its continuous polarity bring the positive on one side to rest itself against, and equilibrate with, the negative on the other side. Such facile continuation of the common polarity would be a properly conducting substance, and all such substances that when deranged in molecular polarity readily recovered themselves, and arranged their molecules by a slight tension urging thereto according to the magnetic meridian, would properly be known as electrical *conductors.* The absence of conductors might leave the electric tension to be permanently resisted, or this tension to so accumulate that it should burst violently through the interrupting obstacle, and restore the equilibrium in a destructive explosion, while the intervention of the conductor might restore the balance by a gradual and silent passage.

13. THE PRINCIPLE OF HEAT.—We have traced the necessary working of the antagonist force according to its own intrinsic nature and constitution, and have found from it the principles respectively of Motion, Sphericity, Gravity, Falling bodies, Magnetism, and Electricity. But the antagonist force does not work alone, the diremptive force perpetually works in and with it, and its generated results keep equal pace with those of the antagonist working through the universal sphere. The diremptive working does not hinder the antagonist activity in the securing of the above principles, but many other results are demanded in a completed nature of things, and which no mere antagonist forces could supply, and which can only be secured in the working of the diremptive force according to its nature and constitution. We now, therefore, fix the insight of the reason directly on the diremptive force working at the centre, and find what must come of it. In it we shall first find the principle of Heat.

While the antagonist force generates a sphere by a perpetual accumulation of concentric spherical layers, the diremptive force will work itself in between all these concentric layers and dispart them by its own peculiar energy. The diremptive activity is in unity with the antagonist activity from the first, and both work on coetaneously, and while the antagonism secures the results we have already traced both unhelped and unhindered by the diremptive working, yet could not the diremptive activity secure any manifested results separate from its combination with the working antagonism. Its action is perpetually away from itself in every limit that might be taken, and this perpetual going off in simple activities, on each side of any limit in which it might be set to work, could only issue in continual self-dissipation, and emptying itself from the limit with no capability of accumulations at the limit. But in working within and against the simple activities of the antagonist force, it at once attains consistency and persistency in space. Its outgoings meet the antagonist incomings, and thereby generate new counteractions, in which there are other determining principles as immutable as any we have yet attained in the ensphering and space-filling antagonisms.

The primitive central antagonism is conceived as two simple activities meeting and counter-working in a common limit, and we now conceive the diremption to be a beginning in that common limit and sending out two simple opposite activities from it. The two mutually and necessarily determine the direction of each other's activity. The antagonism being first, would wake the diremption within it to spread each way in divellency; and the diremption being first, would wake the antagonism on each side of it to push together in counter-agency; and both being in unity of agency, would counterwork each other; and thus in any method of possible communion, the antagonist and diremptive forces must work upon and against each other. From the very nature of the agency in such combination, we may see, moreover, that the result must be a perpetual palpitation, or systole and diastole play between them of a very peculiar and specific kind. Keeping in mind the order of the sphere-forming process, so minutely and carefully followed out in the antagonist working, we will now as care fully trace this process of diremptive working through the whole sphere, and which must necessarily be in the converse order of movement and result, inasmuch as the diremption cannot work and manifest itself at all save in an antithesis to the antagonism.

We fix on the limit in the central antagonist force where act meets act in counteraction, and in that limit we now find also a diremption. The antagonisms pushed each back upon themselves, and made two polar points of force in a line on opposite sides of the limit in which the simple activities had met, and these points, being thus taken and held by the newly engendered forces in them, forbade the antagonisms now to go further back on opposite sides in a line, and by the second principle of motion necessitated a rising out on all sides in an equatorial plane or ring, midway between the polar points and transverse to the axis. Even so, in converse effect, must this diremptive action traverse the same course. In meeting the simple activities of the antagonism and pushing back upon them, it does the same work on each that each one was doing on the other, adding this energy of the diremptive working to the former pushing of the antagonisms, as it disparts itself each way while standing between them. This disparting of the diremptive action in the very limit of the meeting antagonisms lengthens the line in the direction of the polar points, until the static resistance of the polar points equilibrates the diremptive action, and so loosens or disparts the antagonism at its central limit that the diremption must now turn, as the antagonism did, to a direction transverse to its first action. This necessitates a pushing out against the equatorial ring of the first antagonist layer from the central point, and a passing between that layer and the central point, and separating the antagonist layer from the central point by interposing a diremptive layer. This diremptive action then again pushes out the antagonist polar points, both by the central working and the working in the layer without the centre, and this loosens the first antagonist layer in its equatorial ring and passes again transversely through, and against the next layer in its equator.

As then, the equatorial antagonisms must flow down on each side in the meridional lines, and at length rest from all ways upon the polar points, and thus complete their circuit and equilibrate the movement of the antagonism, even so must the central diremptive working pass through the disparted equatorial antagonisms, and down on each side between the outer and inner meridians in the concentric antagonist spherical layers, and occupy the disparted space between the layers in each hemisphere from the equator to the poles. As then, again, the equilibrated movement of the antagonism through the equatorial ring down against the poles has stopped all further passage of the generating force in that direction, and the central working must push the antagonisms out each way again in new polar points, to be resisted by them and turned again through the equatorial plane in new meridional lines, making new concentric spherical layers in each hemisphere; just so must the diremptive force run over again its circuit, and elongate the polar line, and loosen the equatorial antagonisms, and pass through and expand the space between the concentric layers at the equator, and then pass down in each hemisphere disparting the layers to the poles. Thus, while the sphere is engendered and augmented by the perpetual working of the antagonist force, it is also continually disparted and filled in every concentric layer by the diremptive force.

Let it also be carefully noted, that as the diremptive action elongates the axis by pressing back the polar points and filling in the centre by its own force, it makes the sphere to take on a *prolate* form; and that as it passes through the disparted antagonisms at the equator, and presses out against the spherical layer beyond, it pushes out the equatorial region and thus makes the sphere to take on an *oblate* form; and that therefore a perpetual pulsation through every spherical layer must be kept up with every diremptive palpitation of the centre. When the whole sphere is formed, and all its concentric layers filled in with the diremptive forces, if then the diremptive action keep on at the centre, the space between the layers being a plenum, the throb at the centre is felt all through to the circumference instantaneously, except so far as the yielding and elastic spring of the diremptive substance may furnish occasion for a propagated and thus a successive motion. As the diremptive movement passes in the equatorial plane across the margins of the concentric layers of each hemisphere that meet in this plane, the oscillation of the layer must make its vibration of every molecule in the layer, and thus successively of every molecule in the sphere.

Now, this diremptive force is heat in its essential being, and the principle of its working must determine all its phenomenal facts and their laws. In its combination with the antagonist forces it becomes fixed, and goes to the composition of the new substance in a static state, and of course induces no vibrations, and imparts no sensible heat, except as in chemical dissolution it becomes liberated. Its necessity of working in every part of its being is by a disparting and divellent action, on occasion of any force within or against which it may exert its diremptive energy, and thus must *expand* every substance into which it enters in its free state. It may be made to permeate any substance so as only to fill in and occupy all the interstices between and around the molecules, without at all parting or separating the molecules themselves, and as thus held in a quiescent state it will be *latent* heat. It may escape from one substance through the intervening medium, and in its passage through a fluid medium it must follow its order of movement as already given in a vibratory process, and as passing out in vibrations all ways from a thermal source it will give *radiated* heat. While the same substance is radiating its heat, it may also be receiving the radiations from other substances, and such reception of radiated heat will give *absorbed* heat. Some of the radiations meeting a substance that absorbs with difficulty, must pass off from it and give *reflected* heat. The principles determining reflection, refraction, diffraction and polarization are all here included, and may all be exposed in experimental facts, showing that the laws in the facts were all necessitated in an immutable principle. If the diremptive action not only presses between and about the molecules so as to expand the substance, but goes so far as to loosen and thoroughly isolate the molecules that they may freely pass by and over each other, we shall have *thermal fluidity;* a more intense diremptive action, separating the molecules from each other and forcing the particles asunder at some distance, will make *thermal evaporation;* and a greater intensity of heat that breaks up the substantial combination of the molecules themselves, and violently sunders the constitutional forces of the materials, will be *combustion.*

In the universal sphere, the diremptive and antagonist activities will equilibrate, when the working at the centre is just balanced by the reactions from the outlying spherical layers, and the diremptive and antagonist forces run each their own circuit without invading and breaking up one the other. In this state of antagonist and diremptive equilibration we shall have the material universe in that state that may be known as the *primitive ether.*

14. Chemical Principles.—It is not possible that antagonist forces can come into any combinations and thus form chemical compounds by themselves alone. The only manner of their working is by ensphering themselves about a central point, and the entire globe of such forces must be of single and homogeneous molecules throughout. But with the conjoint working of a diremptive force in a converse direction to the antagonist working, there is full occasion given for chemical combinations, and indeed a necessity that such should ultimately be effected. When the diremptive action has diffused its disparting forces between all the layers of the universal sphere, and there is an equilibration of both the antagonist and diremptive forces in their reactions upon the centre, and thus the universal sphere has completed its full destined size, there may still be a continued generation of both these forces at the centre, and instead of augmenting the size of the sphere they will only serve to fill in and increase its density.

When the exact balance is gained, the spherical layers

of the antagonist forces are not only disparted from each other by the infusion of the heat-force in its layers between them, but there is the loosening also of the molecules in the layers of the antagonist forces by the permeating heat-force, so that they just maintain their magnetic polar directions in their meridional lines, and are at the same time just ready to dispart and pass over and by each other in a movable fluid state. The whole mass is just loosened, molecule from molecule, and may be called a fluid, only the molecules are all yet quiescent and not flowing. If, then, the antagonist and diremptive agencies keep on in exact equilibration of their generating activities, there must be a perpetual going out of new forces of both kinds into the universal sphere, but as the antagonist action is just dissolved by the diremptive action, there can be no crowding back of the polar points and extending of the axis by a retorsion of the antagonist simple activities upon themselves, and as there can be secured no extended static in this direction, so there can be no pushing out of the transverse equatorial ring and the successive enlargement of the sphere; and for the same reason there can be no prolate and dilate movement in the going out of the diremptive accumulations, and thus no filling in successively of the heat-force any further between the spherical layers. The consistency of the layers is all dissolved, and thus extension and vibration must wholly cease. And yet generation of both forces goes on, and flows out from the central point into the sphere, and can thus only thicken and make denser the primitive ether without extending its volume. The mass only becomes a thicker fluid, without yet any flow.

But without at all following here the determinations

of such accumulation and thickening of the primitive ether, which will find its more fit opportunity hereafter, it is sufficiently manifest that in such movement and accumulation of forces, there must come new and varied combinations. The antagonist and diremptive activities meet each other, and interwork and dissolve and counterwork each other in many new varieties of action, and we have opened at once all the necessities for chemical affinities, chemical equivalents, and chemical combinations and decompositions.

If one molecule of the antagonist forces be exactly balanced by the working of a diremptive force in the very limit of its simple activities, we have a new chemical atom; entirely a new substance; and competent to stand out alone in complete static individuality. So, if two antagonist forces are just balanced by a diremptive force between them, or a number of antagonist forces around a central diremption, or a varied number of limited diremptive activities by an equilibrating number, intensity, and direction of antagonisms; in all such cases there comes the necessity for all chemical laws. The forces and single activities cannot equilibrate and hold each other in static rest without inducing the whole doctrine of chemical equivalents, and only such forces can run to each other's counteracting help and support as stand in the line of reciprocal activities, and which must introduce the whole doctrine of chemical affinities.

Simple chemical substances will not be single antagonist forces, but such combinations of the antagonist and diremptive activities, in their most simple working, as no application of other combinations can unloose and decompose.

The combining forces must come together in specific proportions, and continue in combination until neutralizing or countervailing forces again destroy their cohesion. The combination cannot be a substance that shall give the qualities of either ingredient, for the combined forces lose all their former mode of activity, and together make a third thing that annihilates the old modes of activity in the new. As the combination wholly suspended and for the time indeed destroyed the elemental forces in their old mode of action, so an analysis of this substance must destroy the force that constituted it, and give occasion again for the elementary activities to work after their former manner.

The perpetuated working at the centre must induce an elementary chaos of prepared forces, that on occasion must come together in chemical combinations.

15. Crystalline Principles.—The free working of the antagonist activities can only form globes which have their single axis; and the chemical compositions that may press their molecules together in a new force, thereby making a new substance, and which may harden it to a body in a rigid state, can only have the one axis which the equipoise of gravity must give to it through its centre. The forces in mere chemical affinities can make combinations only in bipolar and uniaxial bodies, and which poles and axis must be determined by the aggregate of gravitating force, and not by any principle of the combining affinities.

But in the converse activities of the antagonist and diremptive forces, it is plain that there must be occasions for their mutual action and reaction in directly tranverse directions. The diremptive force may stand between and balance two antagonist forces that press together, and this in a transverse direction at right-angles, or at any oblique angle, and such composition of forces must make a nucleus, that in process shall build up around it a cube or a rhombohedron; and if the balancing diremptive force gradually and regularly diminish as the combination goes on, it will necessitate the cutting off of the solid angles of the before-mentioned geometrical solids, and make them to become right angled or rhomboidal octahedrons. Thus may any variety of regular geometrical solids be built up by accordant forces in composition, that shall work towards each other in such directions and degrees as to balance themselves in the axes of such solids. The whole principle of crystallogeny is in this combination of heat with polar matter, or the bringing together of these converse forces in such ratios and directions as will secure multipolar and multiaxial combinations. The whole geometrical solid is determined in its faces and edges and solid angles by the axes and polarity which the working forces secure, and the degrees of energy respectively exerted. And inasmuch as the working forces must also determine the superposition of the facial layers, so the same crystal must always have its permanent determined planes of cleavage, and the specific polish of the faces.

The gravitating force has the principle which determines the fluid rain to fall in spherical drops, and the crystalline principle determines the falling snow to arrange itself in star-shaped needles, and the hail has these congealed in diverse polyhedrons.

16. THE PRINCIPLE OF WORLD-FORMATIONS.—The attained conception of the universal sphere in its generated and accumulating ethereal matter, and the perpetual working of the central antagonist and diremptive activities constantly making new chemical compositions, and thus also new substances in nature, and moving the primitive matter into new forces, must give the occasion for much further tracing the immutable principles to necessary determinations, and without forecasting what we may find we will pass on and see what the clear insight shall disclose.

The circuits of the two primal forces, as they interwork on and with each other, need to be kept distinctly in the apprehension. The antagonist agency goes back in the line of the axis to the poles, and holds itself in static equilibration there until the equatorial ring is elevated transverse to the axis, and until the forces in this ring have crowded their simple activities back on each side upon themselves up to the poles, and balanced the whole movement. Another crowding back of the polar points and lengthening of the axis then occurs, to hold itself again in static rest until the same process is repeated in another equatorial ring, and another hemispherical layer on each side to the poles. Thus perpetually with the circuit of the antagonist force. The diremptive activity, starting from the same limit in the central force, takes the same circuit by a directly converse movement. Going out each way from the central limit, the diremptive activities encounter these antagonist agencies, and thus pushing all back upon the polar points they loosen the central tension, and in this the occasion is given for the diremptive action to turn its divellency transverse to its first direction, and thereby press out and fill in the interval between the first antagonist spherical layer and the central molecule. The diremptive agency thus balanced, must thence again push back upon the next outer polar points in the axis, to be thence turned again transverse to its former direction, and loosen and fill in between the first and second spherical antagonist layer, and thus on alternately prolate and oblate until every spherical layer is loosened by the interposing diremptive force.

This diremptive action ultimately disparts the layers and also the molecules in the layers, and dissolves the whole mass into a fluid or molten state. The two agencies thus balance each other, and the diremption is held still while the antagonism is just parted, and the fluid ether rests quiet.

The principles of these working forces determine clear, though varied and extended results. The polar points are perpetually static, and force the central movements through the equatorial plane and down on each side through the meridional lines to meet in these static poles and rest against them. Each concentric layer is thus balanced in its polar point, and thus each hemisphere throughout is balanced by the aggregate polar points which form its axis, and the two hemispheres stand in antagonism to each other in the equatorial plane, just as the simple activities stand in antagonism at the central point. The diremptive movement, or permeation of heat, is thus ever through the equatorial plane and back each way between the concentric layers, and never through the polar points and out from thence between the concentric layers towards the equator. When the heat is perpetually forced from the centre, as there generated, into the equatorial plane, the hemispherical antagonism tends to hold it in this plane and necessitates its accumulation in the equatorial region. It presses its way between the layers only by overcoming this polar hemispherical resistance. The perpetual generation and effusion must at length isolate every antagonist molecule, and thus truly fuse the whole sphere, but the greatest press of conflicting forces must be in and near to the equatorial plane.

With this pressure of the two hemispheres together in the equatorial plane, from the perpetual working of the central antagonism, and the exact balancing of the molecules in fluid rest by the loosening of the interposed and everywhere permeating heat-force, we have a starting-point for the insight to attain to further determinations. The perpetual pressure of new generated forces, both antagonist and diremptive, into the balanced fluid ether, and which from the loosened and dissolved state of the antagonist forces in the spherical layers, cannot now augment the volume of the universal sphere, must gradually condense and thicken the homogeneous ethereal fluid, and make it to be a chaotic mass of blended and confused interworking forces that, by occasion given, shall come together in chemical combination, and constitute various distinct substances. The even working of the two central forces, while thickening the mass to greater consistency, keeps it still fluid and molten, and ready to flow on any excess of pressure.

This excess must ultimately come, when the consistency of the mass is too dense to permit the ready penetration of the central working forces. A commingled stream of such forces, precluded from free permeation in the thickened chaotic matter, must drive it into currents and force the resisting portions before it into unequal accumulations. The invading currents, meeting with the resistance of the matter in advance, must tend perpetually to spiral and gyrating movements, turning athwart their own courses and revolving across their general lines of movement. Such whirling movement must repeatedly break up the matter that it carries into divers successive separations, and at intervals make wheeling portions of matter that turn themselves about upon their own axles, and work themselves into spherical forms.

These will be of different volume and velocity of movement, according to the forming impulses which have constituted them, but at length the myriads of revolving spheres will have worked up and exhausted the chaotic material, and leave the intervening spaces to be again filled by the purified primitive ether, and the wheeling bodies and their reactions will balance the central forces, and reciprocal regulations and equilibrations must succeed.

The commingled antagonist and diremptive forces from behind, and the hemispherical pressure at the sides, have overcome the forces of gravity and magnetism, and wrought the chemical chaos into these wheeling spheres, but while overworking they have not at all destroyed the forces of gravity and magnetism. These forces have been constant though overborne, and have held the universal sphere steadfast in its own form and proportions. Each new spherical mass has also taken to itself a new centre, and every molecule of the mass is pressed out from the centre according to the principle of repulsion, and pressed back also to the centre according to the principle of attraction, and the determinations of gravity and magnetism are in all these new spherical bodies as in the great universal sphere. Each has the gravity of its own molecules towards its own centre, and each gravitates towards the others according to the ratios of the quantity of matter and inversely as the squares of their distances, and all feel and obey the determining principles of the great universal centre, and arrange themselves in their respective places about it according to the forces in the primitive universal ether in which they have their position. Each new spherical body is also a magnet with its own bi-polar directions, dip, and attraction and repulsion, and yet these must conform to the magnetic forces of the universal sphere, according as the particular new spherical body is in the austral or boreal hemisphere of the universe.

The confused and chaotic mass of chemically combining matter is thus, from the working of its own forces, taking on definite forms and assuming fixed positions, and while made to be individual wholes in their own constitution, they are yet but the component portions of the universal whole which is now coming in to the order of a systematic arrangement. The original antagonist and diremptive activities cannot work on together, without the necessity of inducing such a chaos of chemical combinations, and the necessity also of there ultimately coming out from it this growing order of a systematic whole in its determined and regulated members. All the spherical masses are taken up from the one molten universal mass, and though modified in their places from some peculiarities of forces and affinities, which may give some characteristic chemical differences in the matter of the different spheres, yet must all be substantially of the same material elements and bodily constitution. There is more than analogy in appearance and origin from the same Creator, even an existence in the same primal forces, and a determined nature in the same eternal principles.

We follow these determining principles still onward, and shall find yet more extended results of cosmical order and harmony.

Single Worlds.—When any rotating mass shall be of so great consistency, or of such slow motion, that the revolving force at the circumference is less than the force of gravity, or adhesion, then can no part be separated from the mass in its revolutions, but the forces of gravity and revolution will work on together, and the surrounding fluid matter will be taken up and incorporated into the body, and the wheeling mass must ultimately settle itself into a globe of such an oblate form as the exact compounding of the gravitating and revolving forces shall determine. It must henceforth fill its own place, and rotate alone on its own axle, and take that position in the great sphere which is determined in its specific gravity. Its own forces of magnetism and electricity must be inherent, and it must attract and be attracted according to the universal law of gravity.

Double Worlds.—Should any rotating mass break and separate into two portions of no great inequality, the smaller must revolve outside of and around the larger, as having been in the revolution of the whole thrown off at parting beyond the larger, and while the larger portion must perpetually turn on its own centre, the two must thus be reciprocally yoked together and henceforth remain double. When such double worlds shall be viewed from other worlds in the direction of the plane of their orbits, they must appear to approach and recede alternately from each other; and if viewed perpendicularly to their orbital plane, they must be seen to keep the same distance from each other while they revolve around their centre; and if viewed in any of the intermediate directions, they must be seen to pass from side to side alternately, and always in such conjunction as appropriately to be known as double stars. Such connections in a common centre of revolution must depend upon conditions that cannot be anticipated as of common occurrence, and yet amid millions of forming worlds, the aggregate of double stars in their determining conditions may well be very considerable.

These double worlds must, like the single suns, gravitate together towards the great centre, according to their quantity of matter, and must thus find, and then permanently keep, their proper places in the universal sphere, and stand forever balanced in the position determined by the compounding of the central and all outlying attractions.

Systems of Worlds.—As these spherical bodies rotate on their axes, there must be not only the hemispherical pressure, but the force of rotation, perpetually tending to flatten the spheres at their poles and elevate them at their equators. If any of them have too little adhesiveness, or such an excess of tangential force in rapidity of rotation as to overmatch the force of gravity, then cannot the superficial and equatorial portion maintain its connection with the mass, but must commence its discession at the place of the excess of the tangential force. This disceding portion will be followed by so much of the fluid equatorial circumference as shall at the time bring the rotating and gravitating forces in equilibration, when the sphere will continue its rotations, and the detached portion must follow out its separate determinations.

If the tangential impulse has added but little to the momentum which the now detached portion had when in connection with the spherical body, then must the parted mass move nearly in the old circular track that it had when in the equatorial surface of the body it has left. But if it has a large excess of revolving over the gravitating force, and yet not sufficient to carry it beyond the attracting force of the sphere, the impulse must, proportioned to this excess, carry the detached portion out beyond its old track in the equatorial circumference it has left, and must deviate perpetually wider and wider from it, up to a certain point. The tangential force is constant, and as the detached portion moves off, the gravitating force is perpetually diminishing in the ratio of the square of the distance, it must therefore discede from its old path, till it comes to its culminating point in the opposite end of a line drawn through the centre of the sphere from the point where it disceded from the sphere. From this point, the gravitating force begins to augment and bring the course gradually nearer to its old track, till in the point from whence it disceded its orbit will have been completed, and it must henceforth continue to move through these superior and inferior apsides, in an elliptical orbit around its old parent sphere, and be known as a distinct *planet*.

The eccentricity of the planet's orbit must be directly as the excess of the tangential impulse at the time of discession, for this excess must equilibrate itself in alternate departures and approaches with reference to its old path in the equatorial surface of the sphere. Its inclination of

13

orbit to the plane of the old path must be determined by any forces drawing it aside, or changing the plane of the sphere's revolution after the planet's discession.

This planetary mass thrown from the wheeling sphere, and henceforth to revolve in an elliptical orbit of greater or less eccentricity about the centre of the old sphere as one of its foci, is at the time of its discession in an utterly amorphous condition, both of outward shape and inward constitution and arrangement. It is only so much chemically chaotic matter put in motion in a promiscuous aggregation. But the principle in the determining forces at work, must bring order out of confusion. As altogether separated from the parent sphere, it must have now its own centre of gravity, and every molecule of the fused matter must come under the conditions of the central forces, and tend at once to an arrangement equably about the centre in a globular form. This globular arrangement from the inner central force is also greatly favored by a combination of outward forces. The method of expulsion from the spherical body, and its subsequent action upon the planetary mass, tends strongly to its spherical arrangement and rotation. When the tangential force that expelled it has been in considerable excess, it has necessarily given a proportionally strong impulse to the ejected planetary matter, and this impulse must be the most energetic upon the superior portion of this matter, and tending to drive this forward and over the inferior portion, at the same time the breaking up of the adhesion on leaving the mass, and the perpetuated attraction act most energetically upon the inferior portion, and both tend to restrain and slacken its motion. The combination of these forces necessitates that the upper portion shall run around and spirally enwrap the lower portion. An axis to the planet must thus be generated directly after its separation, and the planet immediately begin to rotate upon it. The direction of its rotation must be determined by these generating forces, and which secure that it must be in the same line in which the body at the time is moving in its orbit. The superior portion must be pushed over, and the inferior portion held back, and the plane of rotation must be determined by such combination. All parts of the planetary mass being of equal consistency, and at first equally distant from its own centre, would give conditions determining that the axis of rotation must be at right angles to the orbit, and the planet's equator in the plane of the orbit, and thus the rotation in the same plane with the revolution.

But any modification of these conditions will vary the determined result. The general tendency must doubtless be to such direction of rotation, but a greater density or a larger volume on one side of the planetary mass must modify its rotation, and come in combination with the other forces to determine where the axis shall be generated, and how one portion shall roll over another. In extreme cases of unequal balance in the planetary matter, the axis will necessarily have an extreme degree of inclination to the orbit. The rotation must ever be conformable to, and never retrograde from the revolution, except from outward interfering forces, but the conditions determining the rotation may give very varied degrees of axial inclination.

The position of the axis and the rate of rotation, being determined by the conditions which first form the planet to a globe, they must henceforth continue the same. Other causes may vary the velocity of revolution in different portions of the orbit, and other forces may come in to change the plane of the orbit itself, but the direction and velocity of rotation are settled in the planet's first formation. The tangential force and the power of attraction in the primal sphere, may vary relatively at different times in the revolution, but the compounded forces of rotation did their work at once, and that impulse is to be henceforth constant in the ejected planet.

When, now, we take the planet as a body rotating on its axis, we can see that, in its fluid state, similar conditions may give similar determinations to it, as were those in the case of the wheeling sphere from whence the planet was separated. This planet rotates with a rapidity determined by the first compounded impulse, and this may be in very varied degrees among different planets. When the excess of force is on the side of the attraction, the tangential force of rotation will make no separations, and the planet will revolve in its orbit alone without any attendant. But when the excess of force is on the side of the tangential impulse, a separation of a portion of the circumference of the planet must ensue, and this portion must form itself into a globe, and revolve about the planet as a satellite, according to the determining principles before given for the planet itself.

The rotation of the planet on its axis will not give the amount of tangential force that the old spherical mass did in throwing off the planet; it is therefore hardly to be anticipated, that when a satellite is formed, it should be made to rotate about an axis generated within. Instead of the tangential force crowding the superior portion on, and the gravitating force holding the inferior portion back, sufficiently to secure a rotation, the presumption would be that when a satellite is formed, it will simply be separated and lifted from the rotating planet, and thus left to concentrate in a globular form, by the action of gravity within it, and merely revolve around the planet. This must give a peculiarity of phase to the satellite in reference to the planet. In the rotation of the planet about its own axis, it abolished the virtual motion that tended to keep its revolution as it had been in unbroken connection with the radii from the great centre, and balanced itself, by its rotary motion, in the plane passing through its centre perpendicular to its axis. Its axis, in each part of its orbit, thus kept itself parallel to the positions it had occupied in every other part, and every revolution turned each portion of the planet's surface, in succession, once towards the great body within its orbit. But the satellite, which does not rotate, has the virtual motion which the fixed radii had communicated from the centre of the planet to the equatorial circumference, and which is now as if the radii were combined in the one radius from the centre of the planet through the centre of the satellite. The satellite, thus, cannot hold itself in any plane passing through its centre perpendicular to some diameter that might be made an axis, but must move on in its orbit, as its parts in their places in the equatorial circumference of the planet had done before their expulsion, when they were fixed in their radii toward the centre. The satellite now separated from it may revolve faster or slower than it, but this satellite must keep on in its revolution with the velocity it had when it left the planet, and with its unvarying phase turned toward the planet in all its revolution. If a satellite rotate, it must turn each part of itself toward the planet with every revolution about it, and if it do not rotate on its axis, it cannot, by its own motion, turn any but the same side towards every part of the planet, as it moves above and around it.

It may also be an occurrence sometimes given, yet seldom repeated, because its conditions can very seldom be found, that the circumference shall be so homogeneous and equable in density, and the tangential force so evenly distributed about the equatorial surface of the planet, that it shall lift a portion of its raised equatorial circumference at once and together from itself, and make a separating space in every part between its own body and a ring above it. In such an occurrence, the separated ring is a satellite which cannot concentrate into a globe. It may condense itself more and more, and the planet also may condense beneath it indefinitely, making the spacial distance between proportionally large, yet unless the ring have force applied in some part to sunder its adhesion, it must perpetually encircle the planet it has parted from, and continue to revolve above and about it. It is itself its own orbit, and each portion of the ring follows every other portion. Its rate of revolution is determined in the momentum possessed at the time of separation, and any changes of generating forces above or beneath must shift its centre of revolution accordingly. In any considerable changes of the force of gravity in or upon itself unequably, there may be a conformity consistently with its integrity while it is a fluid; but as an unyielding dense body, it must be ruptured or precipitated upon the planet, if very powerful or violent disturbing forces act upon it.

Thus may we very clearly follow the determining principles in the formation of world-systems, through their first stages. The oldest planet will be that thrown off from the outmost periphery of the wheeling sphere, occasioned by the compound action of the central and hemispherical forces meeting in the equatorial region of the primitive universal globe of matter. This planet must conform, in its future ongoing, to the conditioning principles we have been tracing by the insight of reason. We have only to follow the same guide, and read the eternal principles in their grounds, to the completion of the idea in the perfected world-system.

When one planetary portion of matter has been thus thrown off from the wheeling sphere, that portion which remains entire must now move on with accelerated velocity. There is both an accumulating force at the centre from the continuance of the perpetually generated pressure, and the parting with a heavy encumbrance from the circumference. Still it must require an action through a considerable period before the now diminished equatorial circumference, though revolving on its centre more rapidly, shall pass in any of its points through an equal space in the same time, that the points in the old circumference did. The tangential force that shall throw off another planet, must require some time in generating. But the point must ultimately be reached, determined by the same conditions as before in the consistency of the matter and the energy of the expulsive power, when another portion of planetary matter must be expelled, and which must go off to work its results under its conditions as before considered. Doubtless its quantity of matter in volume and density, will greatly differ from the former; and also, that there will be a considerable difference in velocity of revolution, eccentricity and inclination of plane in the orbit, rapidity and period of rotation on its axis, and inclination of axis to the orbit, and also a difference in having no satellite or a varied number of them from the former planet. But all these will be determined for it in the conditions that come with it, and these conditions are all given in the determining principles of the great primal forces working under the control of the Absolute Reason at the centre. Thus on, in succession, till the remaining portion of the sphere may be of a density and velocity of revolution, in the exhausted and used-up fused material that was in its gathering reach, that shall permit itself to condense and concentrate in a globe at the centre, and revolve in its own place according to its given conditions, incessantly and interminably, but at the pleasure of the Creator. Each planet and each satellite has settled its own laws of working in its formation, and in them there is a register of the forces and movements of the whole in the places and times of their formation, and the whole is now a unit in its reciprocal interactions, and held also in unity to the great primitive globe which fixes its place for it within itself.

Thus with all single and double and compound worlds; the great amount of fused and prepared material, about the middle regions of the universal globe of matter, may be made to exhaust itself in any number and variety of world-formations, by a directing agency at the centre of all operation, and all these to take their respective places, and revolve therein with unbroken order. When the newly compounded matter shall have been thus all pressed into separate worlds and systems, and the superabundant heat-forces shall be absorbed in such creations, the great globe of universal primitive matter will stand forth, clear in its appropriate forces as before, and the solid worlds be left floating in the pure ether, which has a power of gravity to regulate, but not a consistency and resistance to hinder or derange, their exact revolutions and rotations. The original antagonism at the centre has never been relaxed, and now holds the primal matter, and all the chemically combined matter of the moving worlds, in that energetic force and action which keeps up for them perpetual and palpable existence, shape, and movement. And the conversely acting diremptive-force keeps also its central outgoings constant, and the incessant heat-generation permeates the entire area of the universal sphere. These original agencies now perpetually energize, not that they may constitute new materials, and augment the existing creation, but that they may sustain, equilibrate, and supply the universe in all its parts and uses as already wisely constituted.

The past history of world-formations may be read exactly in their present movements and localities. The central sun once joined in continuous matter itself, through all the intervening worlds, to the outermost planet. These worlds now condensed in solid bodies, were then fluid masses, and the rotating motion of the whole had then at the equatorial circumference, the velocity of this furthest planet in its orbit. In successive stages it has thrown off its superficial strata, which have rolled and hardened into worlds, now moving in their orbits at the same rates of revolution in which the primal sphere was then turning on its axle. Yea, all the universe of suns and systems were once in that chaotic mass together, which was by the primitive heat dissolved, and mingled, and chemically reconstituted out of the modified ethereal matter. They have been separately wrought in their forms, and pressed to their present dimensions, by the strong impulses of the diremptive forces, and the antagonisms which have come together in each hemisphere at the equator. We have only to follow back the record which their facts will bear upon them, and we may read their historical epochs just in the same order of existence, as the eternal principles in the reason would have prophesied their development. The Absolute Creator, with the archetypes of all possible forces in their unmade principles in his own insight, has seen the end that in knowing himself he has known was the most worthy of him, and for his own excellency's sake he adopted it, and for his glory also he has taken that out of all possibles which the principles demanded, that in the highest wisdom he might consummate it. The facts have all been after their unchanging principles; and the benevolence and righteousness of the end has not been by any arbitrary constituting of principles, but by an orderly constituting of facts, and bringing into existence a material creation with a nature that unmade and eternal principles determined for it. The glory of the Maker is, that the making has eternal reasons for it.

17. Principles of Planetary Motion.—The principle determining the formation of world-systems includes within it that also for planetary motion, and we may here follow

out such determination, and attain the three celebrated Kepler's Laws, not as mere facts found and which have no higher explication, but which must so be if the system itself shall be.

(1.) *Planets must revolve in elliptical orbits.* The rotation of the primitive spheres, before the conjoint forces from the centre through the equatorial plane and between the two hemispherical pressures, must secure a translation of the superficial portion, more or less according to degree of adhesion and force of revolution, from the polar regions towards the equator, and thus elevate by so much the equatorial portion of the sphere. While the rate of revolution gives a tangential force less than the gravitating or adhesive force at the equatorial surface, the whole mass must cohere and all rotate about one axis. But when the tangential force at the equator exceeds the gravitating force, there must come a disruption of a portion from the equatorial surface. This may include a thicker or thinner rim of the same consistency of substance, from the conditions of the hemispherical pressure or an interference of external planetary attraction, but the force of revolution which throws off the planetary portion must, to the extent of its excess over cohesion, avail to project it beyond the circular track it had been describing in the equatorial circumference. By the force of its ejection it must pass out beyond its old revolution.

The excess of projectile above that of adhesive force may be to any given amount, but inasmuch as with the greatest degree of projectile impulse, the gravitating force toward the centre of the sphere must still act, so the ejected mass must be affected by it, and cannot pass off in a completely tangential straight line, and must take the course of a curve somewhere within the tangent. If the excess of projectile force be to so great a degree, that when a point taken as a centre within the induced curve shall have lines drawn from this centre to the curve, and then reflected from the curve at the same angle to a tangent at that point, on the other side, which the incident line had with the tangent on this side, and these reflected lines shall also meet a line drawn perpendicularly to the axis of the curve at an angle greater than a right angle, then will that curve be thereby evinced to be a *hyperbola*, and the planetary portion cannot revolve in a complete orbit about the centre. If these reflected lines meet the perpendicular to the axis at an angle equal to a right angle, then will that curve be thereby evinced to be a *parabola*, and the planetary mass still cannot make a complete revolution. But if these reflected lines meet the perpendicular to the axis at an angle less than a right angle, then must they converge and meet somewhere in another point that shall be another focus to the curve, and thus make the curve to have as it were two centres, which will thereby evince that the curve is an *ellipse*, and thus, as returning again into itself, the path will be a complete orbit, and the planet will perpetually revolve in it. Should the reflected lines come back in the incident lines to the same centre, this would thereby evince the curve to be an arc of a *circle*, and such must of course be the orbit of the moving body in it.

But the impossibility that the projected mass should take the curve that is the arc of a circle is manifest in what has already been seen, viz., that when the projectile force was less than, or only equal to, the attractive force, it could not be ejected from the main sphere, and would in the equatorial surface of that sphere describe an exact circle; and that the excess of projectile force, which must eject it, must also to the same degree send it out of and beyond the circle it had been hitherto describing. It must therefore extend its curve beyond a circle if it become a separate planetary body, and it must not extend the curve to the hyperbola or parabola if it revolve at all in a complete orbit. It must therefore take on an elliptical orbit of greater or less eccentricity.

This determined elliptical orbit, and the principle which must determine also the given eccentricity, may be followed in the order of its process. At the point of discession the planet must possess and retain with it a given constant measure of centrifugal force, which, as an excess above the gravitating or centripetal force, has detached it from the primal sphere and sent it beyond the circle in the equatorial circumference. The centripetal force gradually diminishes as this constant centrifugal force carries the planet outwards from the circle it had in the old circumference, and the planet must thus discede continually from the old centre until it has completed one-half of its revolution. But the centrifugal force is not sufficient to carry it by and beyond this culminating point, which would demand that it become a parabola or an hyperbola, and thus the centripetal force avails to bring it down from this point in the semi-revolution, and thence this centripetal force gradually augments through the other half of the revolution to the return in the old point of discession. The inferior apsis must be at the point of discession, and the superior apsis at the opposite point of half a revolution, and the excess of the major over the minor axis must be proportional to the excess of the tangential or projectile impulse. It must be some form of the ellipse, for it must be outside of the old equatorial circle, and it must be inside of either a hyperbolic or parabolic curve which could not return into itself.

(2.) *Each planet must describe equal sectors in equal times in its own orbit.* When the primal sphere from whence the planet was separated rotated with the matter of the planet still adhering to the equatorial surface, each point in every radius in the equatorial plane out of the centre moved in a perfect circle, and each radius from the centre to any point within it, equally distant in all, described equal areas in equal times in the equal velocity of rotation. Since the planet has been thrown off, the projectile force that expelled it has gone with it and remained constant in it, and if the planet had continued to move in its old circular track, the velocity would still have been uniform, and thus its radius or line drawn from its own centre to the centre of revolution would still have described equal sectors in equal times.

But according to the first principle of planetary motion, the excess of tangential over the gravitating force has necessarily given to its course an elliptical orbit of more or less eccentricity, and thus its rate of movement must be variable through all portions of its revolution. This excess of tangential force must, however, exactly balance itself against the gravitating force in the resulting eccentricity of the orbit, and the whole periodic time of revolution must be the same as that of its last rotation in the circumference of the sphere before its ejection. That rotation was in a complete circle, and the radii all described equal sectors in equal times. The radius which the planet now carries with it, or the line from its occupied focus to its own centre, called the *radius vector*, continually lengthens itself in the passage from the inferior to the superior apsis, in the exact proportion inversely as the velocity diminishes; and then again contracts itself in the passage from the superior to the inferior apsis, in its opposite semi-revolution, in the exact proportion inversely as the velocity increases. What is gained in the extent of the radius vector is exactly compensated in the retardation of the movement, and what on the other side of the orbit is lost in the contraction of the radius vector is also exactly compensated in the acceleration of the movement, and the whole periodic time of revolution is the same in the planetary ellipse as it was in the equatorial circle, and thus the circumference of the elliptical orbit is of the same extent as was that of the rotating equatorial circle. But the same extent of radius and arc of the circle have the same proportion to the whole area of the circle, that the like extent of radius vector and arc of the ellipse have to the whole area of the ellipse, and as these are described in equal times in both, and that of the circle is equal sectors in equal times, so that also of the ellipse must be equal sectors in equal times.

(3.) *The squares of the times of revolution must be as the cubes of the mean distances.* If we take a perfectly circular orbit, we may say that a given amount of force will secure that the planet shall have one *minim* of motion in one *moment* of time; and therefore in one minim of motion and one moment of time, one radius will have changed its place, by the revolution, for the place which its next contiguous radius had occupied in the plane of the orbit. But that the same force should carry the planet through its entire orbit, must demand that the same radius take the place successively of all the radii in so many minims of motion and moments of time. And now, as the whole area of the orbital plane is as the square of the radius, so the force is as the square of the minims of motion, and also as the square of the moments of time. A less or greater force, in carrying the planet through the same orbit, must have its proportionally less or greater moments of time, and a less or greater orbit with the same force must have its proportionally less or greater moments of time, and all differences of orbit with differences of force must have their proportional differences of moments of time, and therefore, in all cases, the force and the orbit being given must determine the square of the time of revolution.

Now the orbit may be constituted the same on three different conditions, viz., as the revolution of a line about one of its ends, or the revolution of a circular plane of the same semi-diameter as the length of the line, or the revolution of a solid globe of the same diameter as the circular plane. But while in all these cases the orbits would be the same, in each case the forces must greatly differ one from another. When the planet is thrown from the end of the revolving line, it will move in the same orbit in the same time if the force is as the sum of all the points in the line, or, which is the same thing, as the length of the line. In this case the principle must be, that the squares of the times of revolution must be as the distances.

If the planet be thrown off from the circumference of a rotating circular plane, then also will it move in the same orbit in the same time when the force is as the length of every line from the centre to the circumference, or, as the same thing, the square of the distance. The principle in this case must therefore be, that the squares of the times of revolution must be as the squares of the distances.

But when a planet has been expelled from the equatorial surface of a sphere, although revolving in the same time within the same orbit, yet must its force have been far greater. Every radius of the sphere has thrown off its own portion, and here the principle must be as the cube of the distance, and we shall have the determined formula that the squares of the periodic times will be as the cubes of the distance. The determining forces of the universe in the central antagonist and diremptive working, exclude both the former principles, and give the latter as the third principle of planetary motion.

Thus would it be in all cases of circular orbits, but the principle also equally prevails when the planet has disceded from the primal sphere, and taken on its orbit of a more or less eccentric ellipse. In all cases, the corresponding of centrifugal and centripetal forces must balance the distance from the centre with the rate of motion, and thus always for the entire revolution we must say of the mean distance in the ellipse what is true of the equal radii of the circle, that the squares of the times of revolution are as the cubes of the mean distances.

It is worthy of remark, that planetary formations on either of the three conditions of an expulsion from a line, a circular plane, or a solid sphere, would each determine the same results in the first two principles of planetary motion, and give necessarily elliptical orbits, and equal sectors in equal times, but only planets thrown from the equatorial surface of a globe can make the squares of the periodic times as the cubes of the mean distances.

18. PRINCIPLE OF LIGHT AND OF LUMENIFEROUS BODIES.—Should we conceive that light is some subtle material substance transmitted by radiation from a luminous source, we might have in this that which should impress the organ of vision, but we should have only the substance moved in radiation, and know nothing of the radiating force. Or, should we conceive that this subtle substance was put in rapid and progressive vibration, we should again have that which might make an impression upon the organ, but we should have only the matter vibrating, but not the force which put and keeps it in motion. We need to attain the insight of some force that goes through the midst of the molecules in the primitive ether, and sets them in vibration and registers itself in their movement, before we can know what light is in its essential principle, or determine any of the necessary laws of its phenomena. The diremptive force is the essential principle of light as well as heat. This diremptive activity going out each way in the midst of antagonist forces, necessarily separates and isolates these molecular forces, and in permeating the ensphered mass, it must make its way by giving alternately a prolate and an oblate form to every successive spherical layer, and thereby make every molecule in the layers successively to vibrate as they stand out in their direction and distance from the centre, and thus the vibrations must be a radiation from the centre.

Such diremptive vibrations we have already considered as the principle of heat, and when the vibrations are of a given degree of intensity, breadth, and frequency, they will be essentially mere heat; but this heat intensified to a certain higher degree of sharp and rapid vibration, becomes light. Heat and light are from the one diremptive activity, and are thus one in their essence and general principle, but they differ in tension, breadth, and velocity, and in this difference of degree, will be found all differences of determined phenomenal laws.

We have now the conception of the multiplied worlds and systems, and these as floating, under the control of gravity, in the great ocean of the primitive ether which forms the grand universal sphere. This primitive ether is constituted of the ensphered antagonist molecules, all separated and rendered fluid by the interfusion of the diremptive force between them. The formation of the worlds and systems has collected and conglomerated the chemical elements, and the whole chaotic matter has been thus used up and condensed in separate masses, and the purified ether stands in its own ensphered relation to its one grand centre, and fills up all the interplanetary and interstellary spaces, and stretches itself out to the limits of the universe. The great central antagonist and diremptive activities hold on their steady and equal converse pressures, and thus this sea of the ethereal universe is perpetually tranquil and still. All the diremptive force in it is heat, held in static equilibration by its even surroundings of the antagonist molecular forces, and is thus wholly latent heat. There is much heat in chemical combination in the simple and compound substances of matter, and this heat is also fixed in the matter of the worlds which it helps to constitute, and thus the great amount of diremptive energy is held at rest in its reaction, or rather direct converse action, with other forces. Only the heat, which is free and uncombined in the molten masses of the material worlds, is in a condition to radiate, and so far to make the primitive ether to vibrate.

But in the midst of this general calm of the universe since the progressive settling of the worlds into their forms and places, we now turn the rational insight specially to the new modifications of the antagonist forces in these separate worlds and systems, and we shall attain to some further immutable and eternal principles most interesting, and determining to very broad phenomenal results. So soon as the great chaotic mass of fused chemical materials for world-formations in the universal equatorial region, had been first thickened and then broken up by the crowding in of the augmenting central forces, and the various streams had been rounded into spiral circuits by the advance resistances, and these divided and condensed into separate wheeling spheres, then these separate spheres all had each its own centre, and every molecule in them at once had a polarity that was determined from the particular centre, while only the general polarity of the particular sphere was determined by the centre of the great universal sphere. This particular polarity and gravity for each molecule to its new centre, has been again modified by the expulsion of each new planet and satellite, and which have each had all their molecules turned in polarity and gravity to conform to their own new centres; and yet all the particular globes of the same system have kept the one and same common centre for their general polarity and gravity. And especially, while the worlds of the system have had their one common centre, the ethereal molecules through all the spaces of that system have been turned and kept in their determined polarity and gravity by that centre. The primitive ether within the system has been ensphered about the centre of the system, and we may regard it as a particular ethereal sphere separate altogether from the globes of chemically combined matter that float within it.

Take, then, the ethereal sphere of any particular system, and which extends out from the centre, beyond its furthest planet, to the extent to which its power of attraction reaches, and every molecular force of the antagonist activities is pressed out and presses back, just in proportion to its quantity of force or matter, and inversely to the square of its distance from the centre. Now when the central sun shall have thrown off all its planets, and settled itself in a sphere at the centre, its surface must take and sustain all this ethereal gravitation. The diremptive force, which mingles in it and so permeates it as to surround and isolate every antagonist molecular force, is wholly imponderable; its action is away from the centre, and is held from expansion to infinity only by the antagonist activities which meet and retain it; the antagonisms only press back upon the centre, and thus the antagonist molecules only are ponderable, and all these through the sphere do press upon the sun's surface.

In this sphere of the ethereal fluid surrounding the central sun, we must regard only the antagonist forces as pressing back upon the body of the sun in their gravity, and while the permeating heat-force would just dissolve and make the whole to be fluid ether, if only the general pressure to the universal centre were regarded, yet now with this new centre of the world-system and the pressure of the antagonist forces back upon it, the ether cannot remain a wholly dissolved and every way movable fluid. The polar points, and equatorial rings, and spherical layers of the antagonist forces, all again stiffen into consistency and rigidity, by the gravity pressing back upon the central sun, and the heat-force through all the ethereal sphere of the system is confined between the spherical layers, and thus an action must at once commence at the centre, or on the sun's surface, that will send the diremptive forces in their alternate processes of prolate and dilate movement through the sphere, and on beyond till equalized by the outer resistance, and then the radiations can propagate themselves no further. The new direction of gravity to the centre of the system must at once make a new radiating movement from that centre, and the pressure of the ether upon the sun's surface, and the friction of the sun's rotation in collision with it, must accumulate a large amount of the diremptive force about it, and which will at once be a perpetual source for outgoing streams of vibrating energy. Such diremptive generation and accumulation must be constant, and as the pressure is greatly intensified, so the vibrations are proportionally quickened and sharpened, and the heat becomes light, and the sun has its luminous atmosphere, and is the great centre of heat and light, as well as of gravity to its system. Its light and heat are as determinate principles as its gravity, yea, they are eternally determined in its gravity. Such a centre to a system cannot be in the universal ether, but it will kindle its heat and light about it, and diffuse it abroad in perpetual vibrating radiations.

This gravitating pressure at the centre is constant; and the supply from the great ethereal universal sphere is exhaustless; as the radiations go off to mingle in and be absorbed by the great ethereal ocean, so the great central pressure that makes and keeps that ocean full must work the diremptive forces back to the solar centre, through all the layers of the sphere that its gravity has formed about it, as the radiations go out and exhaust that which is in this centre; and thus the solar light is perpetually replenished, and its radiations in successive vibratory movements continual, and the circuit of heat and light unbroken.

The forces of gravity and the tangential repulsions must determine the rate of revolution, and thus the point at which the central body shall cease to throw off new planets, and must thus also determine the volume of the central sun, and this will regulate its amount of light. Before the last planets shall have been separated, and while the system-making sphere is yet of very considerable size, the forces of gravity in the ethereal fluid will begin to press upon its surface in sufficient intensity to wake and actuate the latent heat, and commence the faint accumulations of a luminous atmosphere, and which must grow on as the volume of the central body diminishes and the forces of gravity upon its surface augment, and thus light will have been generated, and day and night have been given to the planets already thrown off, before the sun as the permanent centre shall have been constituted. The morning twilight of creation must begin in the first kindling of light upon the surface of the diminishing system-sphere, and become the perfect day when this system-sphere has become the central sun, and its full light reflected upon the planetary bodies shall make it and them to be "lights in the firmament of heaven, to divide the day from the night, and to be for signs and for seasons and for days and years." Were the central sun to diminish less than this due proportion determined by gravity and heat-force, both light and heat would become deficient. Every central sun must be such a luminiferous body, and while a sun to its own system, it must also be a star shining in its own light to all the other suns and worlds, within the scope of its radiating vibrations. The planets will not be centres of gravity, giving intensity to the antagonist forces in awakening the latent heat sufficiently to make them self-luminous, though it may so be that some of the larger shall have their light increased by this self-production beyond the amount of reflection.

The whole sphere of the system, and indeed the universal sphere of the primitive ether, is filled with static or latent heat, and thus the central radiation has not to pass in locomotion all through the interior and out to the circumference in its vibrations to propagate light; the pulsation at the centre finds a plenum before it, and thus each throb moves the whole, and only the compressibility and elastic spring is to be estimated in the transmission of motion. The rapidity of light will be uniform and in almost inconceivable degrees.

As radiating from a centre and thus diminishing its tension in the ratio of gravity, it must give its degree of intensity inversely to the extent of the subtending angular line, or apparent size of the body, and thus degree of light and size will diminish alike by distance, each being inversely to the square of the distance. A small body of proportionally intense light may appear equal in brightness, at the same distance, with a large body and less light.

The principle of radiation in the vibrating movement from a centre being attained, it will at once determine what must be the phenomena. If the transmitted vibrations meet a substance that stops and absorbs them, such interposing substance must be *opaque.* If they readily pass through the substance, it must be *transparent;* and if passing scantily and with difficulty, it must be merely *translucent.* If they strike some substance and are deflected in their course, they must thereby determine all the phenomena in *reflection;* and in passing from a rarer into a denser medium, there must be *refraction;* and if passing by the opaque edge of an interposing body, there must be *diffraction.* If the substance transmit the opposite vibrations unequally, there must be *double refraction;* and if a direct ray be refracted unequally, it must in that process be analyzed and spread its unequal vibrations through an elongated *spectrum.* Also, if a direct ray be reflected or refracted at a particular angle, and then be turned on its axis, it will have made one side of the vibrations to disappear, in turning the edges of their plane within the line of vision, and only the vibrations in the transverse plane can be apparent, and thus the radiation must be *polarized;* and if one line of vibrations cross another, there must be alternate combination and neutralization, and thus the phenomena of *interference.* The principle being attained, the laws in the facts are a necessity.

19. THE PRINCIPLES OF GEOLOGICAL FORMATIONS.—All worlds are constituted from some combinations of the two elemental forces, and while the circumstances may somewhat modify the chemical compounding, and vary in different worlds their material substances to a certain extent, yet must each world be essentially like the others, and the combination and collocation of its substances be peculiar only in the peculiar conditions to which it has been subjected. We may thus apprehend the general principles of this world-formation, and some of the peculiarities which certain of the worlds from their circumstances must possess, and we can in this have the principles of what may be termed a universal geology.

The universe must have its particular systems, and the system must have its particular sun, planets, and satellites. All must have the same determinate principles of planetary revolution. The planets may condense in the cooling and conglomerating process unequally, but in the original throwing off, the outer planets must have been the rarer and the inner planets the denser, and such must continue to be their general state in their perpetual orbital positions. The central suns, from their perpetual atmosphere of heat and light, must condense the least from cooling. The particular world as a planet, will have the action of gravity concentrating all its matter by direct radial lines to the centre, and added thereto a tangential force in its rotation about its axis. This general action of gravity will bring the denser matter to the centre, and the whole outlying matter will press in upon it, and thus secure that the matter in the first fluid formation will be ranged according to the universal law of a density in the inverse ratio to the square of the distance from the centre. But a rotary force will so modify this, that in addition to the direct concentration of gravity there will be the whirling movement, wrapping layer over layer in concentric folds. The globe will thus form in concentric strata of perpetually increasing density to the centre, and with a tendency, when the mass may harden, to a cleavage in the direction between the strata, and to lines and fractures in the direction of gravity across the strata.

The cooling process is the escaping of the superabundant and uncombined diremptive forces, which go off from the very nature and constitution of their existence, leaving the chemical combinations in the body to their unhindered strength of affinities, and thus the substance of the planetary world must become at length a solid globe of less or greater volume and density. As the cooling and condensing process goes on, the crust above the fused matter becomes thicker, the outer strata press their weight upon the lower, and therefore from both the necessity of the heaviest being the lowest in a fluid state, and the outer pressing the inner in a solid state, the lower strata in position must be the densest and hardest, and the most compact crystals and sub-crystalline rocks will have their places nearest to the internal fire. The first geological formations must be plutonic, the crystallized and partially crystallized will underlie the composite, and the inner heat will at length be so confined and softened, that an atmosphere shall form, and the combination of water commence, and ultimately the wernerian geological process must begin. Disintegration and abrasion, and diluvial currents, will make their transpositions of substance, and sedimentary deposits will make their various strata, and vegetable and animal life begin, and their fossil remains become imbedded in the forming and successively overlying portions. The sea and air will alternate over the same places, and each make their distinct record upon the same portions of the planet's surface, and leave it to be read by future philosophical observers. The force of the inner fires, especially in any commingling with the expansible fluids and gases, must induce wide disruptions and upheavals, and the tilting and twisting of the superincumbent strata, and thus the surface of the planet must become broken into ridges and fissures, chasms and opening caves beneath the solid overlying portions; and in the greater labors of the subterranean forces, mountains and valleys must be formed, and broad fields of the horizontal strata will be upturned, and give their outcropping edges to reveal the orders of nature's ongoings for long geological cycles.

These inner fires must often have their orifices opening upwards to the surface, and the volcanic action from these open craters will give the index of the disturbances beneath; and when obstructions to the volcanic vent occurs, or new explosions take place under the solid crust, there must come in connection all the violent tremblings and commotion of the earthquake. Continents will be lifted or depressed, and oceans and lakes will swell or subside, and the surface necessarily take on all the modifications given to it by the movements of the fused mass on which this superficial crust reposes.

20. The Principle of Cometary Bodies.—When the fluid and variously compounded matter, in its chaotic

state, has been very generally taken up and wrought into revolving suns and systems by the pressure of the central forces, there must still be a residuum which has worked itself off from, or which has never been taken up by the wheeling worlds in their forming state, and which must be subject to the determination of these forces which are still working in the midst of it. Its combination chemically with the heat-force has given to it a different composition from the primitive ether, and a greater density and consistency, which forbid that it should blend and mingle with this ethereal matter now surrounding the ensphered worlds; it must thus concentrate itself into many detached spherical bodies, and which must be of much greater rarity than the suns and planetary systems. These varied nebulous globes must be moved by the impulses of the world-forming forces, yet in constant though diminished activity, and must also feel the attractions of any world to which they make an approach. As the universal ether thus clears itself by these nebulous condensations, an indefinite number of such râre bodies, larger and smaller, must be floating between the worlds.

While thus flitting amid the open spaces, it must not unfrequently occur that some of these rare bodies will come within the attractions of some of the dense systems. The way for an entrance is open from any quarter, and the rate of velocity may be as various as the compound impulses which urge them on, but so soon as they come within the gravitating influence of any system, they must be subject to laws that we can fully estimate, and we may very fully apprehend the results that must be determined for them. As they come within any system, they must move towards, and ultimately fall into, or pass around, the central body of the system. In thus passing about the centre, the nebulous body must take on one of the following forms in the track which it makes; either a hyperbola, a parabola, or an ellipse, the last of which may have so little eccentricity as to approach the form of a circle.

When the compounded impulse and attraction shall give the hyperbolic path, the body must pass and recede from the central sun in a diverging course from that by which it entered, and as it leaves the system on an opposite side from its entrance, its track must be perpetually divergent, and the action of its present forces can never lead it back again within the same system. Should the compound agency determine a parabolic course, the entering body must also pass about the centre, and approach towards a parallel direction with the path it entered, and keep on its outward way in a perpetually receding journey. Many may so pass through and leave a particular system, and though ever afterwards modified in direction and velocity by it, yet never again visit it.

But when such a body shall so enter a system that the momentum it brings, and the attractions it receives, shall shape its track about the sun to an ellipse, it is then caught by the system, and must henceforth abide within it, unless some subsequent acceleration or retardation should induce its movement to one of the former curves. It may be anticipated that more will pass through a given system than will be caught and retained by it, but that many floating nebulæ must so be arrested by particular planetary systems, the known conditions are too favorable for such a result to permit that the facts should be doubted.

Such as enter more nearly in the plane that conforms with the general planetary orbits, will meet the most interfering forces, and be most likely to be kept within the system, and any particular system will have the greater probability that the larger number of such bodies which it takes into its company will be in orbits that are in the general plane of its planetary bodies. The conditions admit, however, that such bodies may become incorporated with the system from any quarter, and that the movements may be either direct or retrograde compared with the revolving planets. The impulse that brings the body in must determine in composition with the attractions that the body finds, what eccentricity shall be given to the orbit, but the occasions are open for such orbits as shall be nearly conformed to the planets, or such as shall in extreme eccentricity bring their perihelion distance close upon the margin of the sun. Some may have their orbits far within the system; some only just within the orbits of the outer planets; and some may stretch their orbits far beyond any circling world the system knows. All such bodies, occasional or constant, are properly comets.

21. The Principle of Stellar Distribution.—A careful insight will determine from the working of the central forces how the stars must arrange themselves, and what the shapes and localities of the stellar fields, as truly as how the planets must be arranged in their respective systems, and what shapes the planets and their orbits must assume. The principle is of broader application and controls over the universal sphere, but it is as thoroughly intelligible and readily explicable as the principles which determine the bodies and their revolutions in the particular planetary

system. We only need the corresponding care and patience in the investigation.

When the two central forces work in unison, the antagonisms form and maintain in constant tension the spherical layers that enclose the centre, and which have their mutual counteracting pressure in their equators, and the diremptions pass out in the equatorial plane and down between these spherical layers to the poles, and keep the layers separate by this heat-force between them. If the diremptive action become an exact balance to the antagonist action, these spherical layers become not only separated by the heat-force between them, but the heat-force permeates the layers themselves and just dissolves the molecular forces that constitute them, and thus the whole internal structure of the universal sphere becomes separable in all its molecules on every side, and is thoroughly a fluid.

In this fluid but still quiescent because equally balanced state, the antagonist spherical layers and the diremptive separating forces between them all equally counteract, from the opposite poles, in the equatorial plane, and the universal sphere becomes a composite of two homogeneous hemispheres, that antagonize altogether with each other in the equatorial plane, just as the semi-diameters from the poles antagonize with each other in the centre. A diremptive action, therefore, now going on at the centre and pushing out its divellent heat-forces into the universal sphere, would not go through the alternate processes of polar prolation and oblation, as before this universal dissolution and fluidity of the molecular forces. The spherical layers being dissolved they could not each hold the diremption, first in the static polar point and turning it thereby to a transverse action in the equatorial plane, and then turning the diremptive force down between the next contiguous layers to fix another static polar point exterior to the first, and so on through all the layers successively, making a perpetual radiating vibration; but the diremptive action, instead of taking this alternate leaping process from layer to layer, will now be able to pass equably right on through the dissolved layers, just where the compounding of the forces acting shall determine the direction of the movement.

And precisely the direction which such diremptive movement must take is the one now to be determined and followed by the insight, for this going out of the perpetually generated central forces, both antagonist and diremptive in conjunction, into the fluid molecules of the sphere, and thereby more densely filling up the ethereal matter, is the very process by which the chemically chaotic matter as elementary for the planetary systems with their central suns is to be constituted. The determining where in the universal sphere this chemical material is to be formed, must also be the determining where the systems and their suns must be that shall be formed out of it.

We start, therefore, in this further investigation, with the universal matter in a perfectly fluid and quiescent state, and with the perpetual generation of the antagonist and diremptive forces still going on at the centre, and their conjoined stream forcing itself out and permeating all through the ethereal fluid, and we must look at the composition of forces here at work, to determine where this thickening and augmenting world-material must arrange itself, and become pressed into revolving suns and systems. The two hemispheres now antagonize and hold each other in a state of rest, by their balanced counteragency in their line of contact in the equatorial plane, and their whole fluid content is homogeneous. How then must the augmentations of the conjoined antagonist and diremptive forces at the centre, distribute and arrange themselves?

The hemispherical pressure is generated in the reactions of the central antagonism, and in the aggregate must be as the cubes of the axes of the hemispheres. The greatest pressure, at any point in the equatorial plane, must be at the centre, where the hemispherical axes meet and counterwork each other; and any point in the equatorial plane out of the centre must have its pressure inversely as the cube of its distance from the centre. So, also, the conjoined forces generated at the centre, and which are to go out against the hemispherical pressure and permeate the ethereal matter of the universal sphere, have their greatest energy at the centre, and any points in the equatorial plane will also have the conjoined forces in them inversely as the cubes of their distances from the centre. The conjoined generated forces at the centre, so soon as they rise to any excess above the hemispherical pressure, must move out into the ethereal fluid matter of the sphere according to the determinations of the compounding of these forces, viz., the conjunct central antagonist and diremptive forces crowding out and the hemispherical pressure pushing in, and these will be equal at the centre, and of equal ratios at the same distances from the centre. We have only to follow such determinations, and the distribution of the world-material must be given.

But this composition of the hemispherical pressure and the crowding out of the central conjoint forces is so complicated, that we here first find the expediency of referring to a diagram, and will call to our aid the representations in the accompanying figure.

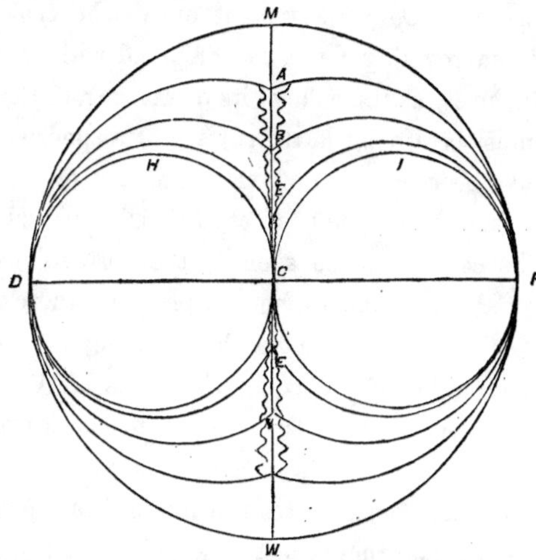

Let M F W D be the bisection of a sphere through its polar diameter D F, and the line of the equatorial plane is then M W. The hemispherical pressure is thus from the poles D and F and towards the centre C, and holding itself statically at rest in the whole line of the hemispherical junction at the equatorial plane M W.

The greatest pressure is at C, and from the direction each way in the hemispherical axes D C and F C. Any point out of the centre will be one point in the circumference of a sphere of points about the centre, and thus have a pressure, compared with the pressure at the centre, inversely as the cube of its distance from the centre.

On the other hand, the conjoint antagonist and diremptive forces, generated at C, crowd outward in the opposite direction against the hemispherical pressure D C and F C, and when at all in excess must move out from the centre C. The greatest force of this outcrowding movement is in the source C, and any point out of the centre must have its outcrowding force, as compared with that at the centre, inversely as the cube of its distance from the centre, since it must be a point in the surface of a sphere of points which have been crowded out from the centre.

Take then the conjoint forces in their perpetual generation at the centre, and so soon as they rise to any excess of energy above the hemispherical pressure, there must be a movement out of the centre, the tendency of which will be in the direction of the equatorial plane M W all about the centre, and perpendicular to the hemispherical pressure at the centre. But as soon as there is an arising out of the centre, the hemispherical pressure diminishes on both the sides D and F, and there must be a parting of the conjoint forces on each side of the equatorial plane, and a compounding of one part with the hemispherical pressure on one side D, and a compounding of the other part with the hemispherical pressure on the other side F, and this composition of forces of equal ratios, inversely as the cubes of their distances from the centre, must make the movements to be an ensphering, by two spherical strata one on each side of the centre, of the compressed ethereal matter into two globes whose diameters shall be the two hemispherical axes D C and F C. Within these globes of the compressed ethereal fluid, the conjoint antagonist and diremptive forces cannot crowd themselves, but must move and form their stratum, of the new elemental chaotic matter they now make, in these two enclosing hollow spheres D H C and F

I C. The onward generation of these conjoint central forces must then make its movement out from the centre C, into the equatorial plane M W above the two spherical strata D H C and F I C, and ensphere them by a superimposed stratum upon each of them; and thence onward again by other superimposed strata on each side, till there shall come, in the growing strata, an equilibration to the central generating energy, and the formations of further strata will then cease.

The superincumbent strata cannot be complete spheres, but perpetually diminishing portions of constantly enlarging spheres, till they come to the universal circumference, where they must be the periphery of the two hemispheres. For, take any point out of the centre in the equatorial plane, as E, and at that point both the hemispherical pressure and the outcrowding conjoint central forces have alike diminished, and are inversely as the cubes of their distance from the centre, and therefore the spherical strata E D E and E F E must be less each than a complete sphere, by the spherical arcs in each whose còrds must respectively be double the distance E C, that is E E. And the same may be shown for any other points beyond E, as B and A, and thus on to the circumference M; where the spherical strata on each side at B must be of a larger sphere than those at E, but a less portion of a complete sphere by the difference of an arc of a sphere whose cord is double the distance B C; and on each side at A a still larger sphere, but less a complete sphere by the spherical arc of a longer cord that is double the distance A C; and then at M the spherical stratum becomes the extent of the universal sphere, but less a complete sphere by the arc of a hemisphere, or an arc whose cord is double the distance M C. The whole thickening of the fluid ether and formation of the chemically chaotic world-material must be out of and beyond the two central globes of compressed ether D H C and F I C, and with the incumbent strata successively of enlarging spheres, but diminishing portions of the spheres till the hemispherical periphery is reached.

When this matter has been made too dense for the conjoint central forces to penetrate it, then must the stream of these forces drive it into whirling spheres, and these spheres into suns and revolving systems, and the general planes of the orbits of the systems must be at right-angles to the tangents of these spherical strata, in which the impulse of the system-forming forces must move. The place for all stars and systems must therefore be in the regions beyond the central hollow globes, or globes of pure compressed ether, D H C and F I C, and within the circuit of the universal sphere M F W D. The stellar strata must be the thickest near the centre, and diminish as they recede towards the universal surface; and the greatest number of systems and their central suns must be in the neighborhood of the equatorial plane, and the stars pretty rapidly diminish in numbers as they stand back from the equatorial plane towards the universal polar regions. Inasmuch also, as the outcrowding currents will not work the chaotic matter into suns and systems, until they have driven it some distance back from the equatorial plane, so there must be a vacancy of stars on each side of the equatorial plane, diverging from the centre as the matter grows thinner, and represented in the figure by the waving lines that fade away toward the confines of the universe.

When the interstellary spaces are again cleared from all but the primitive ether, the central forces, though not accumulating, will still flow through these ethereal seas, and must necessarily waft the floating stars into varied island groups, and their distance and positions must give to a spectator from any one, all the varied phenomena of stellar clusters, and unresolved nebulæ, and changing position without revolving motion.

LIFE.

Up to this point we have followed the generating and arranging forces working in and from the centre, and stupendous as have been the results in the determinations of universal nature, they have still been merely mechanical, and the same operations perpetuated endlessly can never lift themselves above the sphere of matter, nor produce any thing beyond material and mechanical changes in nature. How the universal cosmos may be originated, and how it must then be orderly and harmoniously arranged by the determinations of its central forces, and the wonderful beauty that comes out in the consummated structure, may all be apprehended in the rational process which we have so carefully and extensively pursued. Still the whole, vast and complicated yet orderly as it is, can be nothing but a magnificent machine; its whole substance is the balance of static, and its whole causal energy is the preponderance of dynamic forces. We have the forces in which matter is, and the principles of their working determining what matter does, but all is mechanically pushed or pulled into its shape and proportions.

This mechanism will work on in the worlds, and when the superficial strata have cooled and hardened to a permanent crust that admits collected gases to combine and form themselves into vapors and mist, and these condensing in water, which, as superincumbent upon the solid earth, gathers itself into ocean beds, and then both land and water become enveloped by an atmosphere through which everywhere the radiations of light are reflected and diffused, there then comes an occasion for a higher order of existence than any chemical combinations or crystalline concretions can reach. The eternal archetypes of *organic* being are in the Absolute Reason as a distinct kind of existence, where the one activity is everywhere within itself both means and end, and making the whole to minister to each part as truly as each part ministers to the whole, and such archetypal being must, for the consistency and satisfaction of reason itself, be somehow embodied in objective manifestation and actual realization. It behooves the Absolute Spirit for reason's sake, or which is just the same meaning in other words, for the sake of his own glory, that he superinduce upon the forces now working in nature a higher force, that may take these mechanical forces into its service, and use them without destroying them for its own organific purposes. They can make exact combinations in all chemical substances, and build up layer by layer about a nucleus the geometrical solids of all crystalline bodies, but in all these cases the work goes on solely by accumulation of parts. The least portion of an earth, or metal, or crystal, is a unit as perfect in itself as the aggregation of the largest bodies, and no possible working of such forces in accumulation, can make the whole to be an organism, where no part is a unit without the whole and the unity of the whole depends upon the presence of every part. As well might material nature have originated from the empty void at first, as that now this new and higher form of existence in an organic being should come out of a nature which exists only in mechanical forces. From nothing, nothing comes; out of material mechanism living organism can never arise. Organic existence should be, for reason sees that mechanical forces are incomplete without a superinduction of living forces; the Absolute Spirit cannot approve to himself his own work, nor rest satisfied in the glory of his own being, by stopping in his creating and governing agency with the material; he must put the vital also within the material, and so overrule and use mechanical forces, that while they continue to be still matter, that matter shall no longer be an extrinsic combination, but an intrinsic living assimilation and incorporation.

This new creating work is not now needed at the great centre of the universe. All that is material and mechanical gathers itself for its sustentation and direction immediately back within the great central working sources, but while these central agencies thus uphold and guide all nature's substances and causes, it now needs that the Creator put his hand upon nature, and work his originations in the midst of the material elements that lie prepared upon the surface. That new creation must be such as shall vitalize and organize these material elements, and the task now is to gain so distinct an idea of this vital force, that while it shall fully discriminate itself from all antagonist and diremptive activities, it shall also be a sufficient ground for the insight of reason to see how the great principles of life and organization are determined necessarily and universally from it. As the antagonist force was more plainly read by the reason than the diremptive, so it may be anticipated that the organic forces will be more hidden than the mechanical—the principles of life will have a deeper mystery than those of gravity and magnetism, or even of heat and light, and must be longer studied before they can mature into a completed philosophy.

22. Life an Assimilative Force.—The Antagonist Activity is a force from its own working; it counterworks upon itself at every point, and thus doubles back each way upon itself on each side out of the point of antagonism. The diremptive activity is no force in itself beyond the mere point of diremption; it outworks from itself in every point, and thus discedes and disparts from itself on each side out of the point of the divellent action. Were it to work in a void, it could never fill it, but must perpetually be leaping each way from the points occupied. Diremptive forces, working in and among antagonist forces, become truly determinable forces, for they are held and work in determinate spaces by the antagonisms they are perpetually encountering. There may thus be perpetual solution and combination, resolution and recombination, through all time, and to as great a variety as the arithmetical permutation of given quantities in the directions and degrees of energy shall permit. But the activity we now seek in idea must be one that can use these for its own ends, and while it makes them work for it, it must also work in and through all of them which it uses, so as to make them to become the common members of the one

complete organism. In a word, we must have an agency that can take the material agents to itself, and assimilate them to each other in itself, and so state or posit them in continuation that while it shall be their builder, they shall become its body, and both together constitute an existence that has unity and identity throughout. It must draw that which is without itself into itself, and incorporate it with itself, and thus truly it will organize itself in living matter.

That which takes to itself, or draws in from without, must originate a movement to which the without may be a condition or an occasion, but for which it cannot be a cause. There must be *in*drawing before there can be *ex*hausting; the spontaneity of the organific agency must thus be on the inside. The living force must first act, or the mechanical forces can never become assimilated and incorporated; they might continue to act upon each other, but can never else be made to interpenetrate each other. We take then a simple spiritual activity, of which we can predicate in itself neither extension nor duration, for it has no *where* in order that we might determine place, and no *when* in order that we might determine period; but we put it into the midst of nature's space-filling and time-during forces, and let it register its action in them, and we can determine for it both a space and a time. Itself spiritual, and perpetually in itself maintaining its own simplicity of agency, and thus wholly incognizable by any sense, its working upon the material forces that impress themselves upon the senses, gives its results in matter to become phenomenal, and thus the modifications which the living force makes of matter, may be readily subject to human experience. To the reason's eye, we must therefore subject this spiritual life-force, that we may therein determine the principles of its modification of matter, and the conditions under which it must build its body for any sensible manifestation.

This simple spiritual activity can in itself fix upon no place, nor hold itself in any position, and it may thus be said to have a want, and must necessarily act for its supply, and thus seek some of the material forces against which it may work and balance and sustain itself. Though all unconscious of its wants and of the adaptations in material forces for its supply and relief, yet will the activity go out spontaneously to its appropriate material forces as if it had already a sentient guide and directory. Some of the material forces in nature will be fit for its use, and will readily combine with it, and thus fix it in position by counter-working with it and truly becoming assimilated in it. In thus combining with the life-force, the material force will part with some of its own activities, and become thereby new substance in its assimilation and combination with the living spiritual activity; and leaving also the activities in the forces used, which become liberated in the vital combination, to combine anew with other material forces about them, it will thereby change also the substances in material nature. There will thus be truly a vital chemistry, both in the new living combinations, and in the changed combinations in matter from the unused activities liberated in the vital action, and recombining with the material forces about them. It is, however, only with the combinations immediately made with the vital activity, that we now need to have any dealing. The spiritual activity combines with such material activities as it finds fitted to its want, and thus fixes itself to them and in them, and they become a new substance by being thus taken up in the life-force. Matter and spirit are in this truly blended, and the life-force is no longer merely spiritual activity, and the matter is no longer mere gross mechanism, but this third thing as a mere substance is indifferently, either life embodied or matter vitalized.

This vitalizing and thus assimilating and corporealizing process necessarily makes new voids in the old material forces. The life-force has taken in and thus taken away from their places the forces and activities it has used, and thereby a vacuum at once supervenes between the living corpuscle and the mechanical molecule, and the forces of nature must at once press up and bring new matter to the living action, which is also used in new living combinations, and thus the living body continually feeds upon these new materials and grows by their successive assimilations and incorporations. As an activity combining with the material forces and taking them up into itself, the vital action may be termed an *assimilative force;* and as thus making a void for external nature to pass through up to the working activity, it may be known as an *indrawing* or an *inhaustive force*, and either may hereafter be applied as the particular application may determine to be the most appropriate. The assimilating is truly the work of the life-force, and is first; the void thus made becomes the occasion for the indrawing, and which is truly nothing but the inpressing of nature, and thus the life-force uses nature in two ways, both for its own want in the fixing and stating itself in combination with matter, and in making the material molecules to press themselves through the voids made by former assimilations up to the spiritual activity, and thereby supply new material for further assimilation and growth.

And inasmuch as the life-force works spontaneously for its own wants, so in this must it be an agent working toward ends, and determined in its activity by just the wants and therefore the ends inherent in it. And this must make it a *formative* force, having a *nisus formativus* or form-making principle in the spiritual activity itself. The form in the rain-drop, and in the crystal, is in all the many drops of the shower and in all the many portions of the crystal, but the form in the living body is one already in the life-force, and works itself out and registers itself in the living body. All living forms are thus determined in the specific life-force, and the whole body must be built up as a self-realizing product of the spirit. General resemblances may enable to classify living bodies into their kingdom, class, order, tribe, family, and genus. These classifications rest on extrinsic and contingent relationships, but where the distinction is that of type in the primitive vital force, and which is brought out in manifestation in the progenitor, and is individually carried down through all the descendants, the classification is then of specific differences and not of general resemblances, and is first into different species, these species into different races, and these into different varieties.

When the specific life-force is once embodied in its organized material assimilations, it must keep on ever working in the same body, growing as it extends itself in new combinations, and when these augmented combinations have extended so far as to equalize the assimilative force, and balance the new assimilations only against the old absorptions and exclusions, the body has then come to its adult and mature stature, and while ceasing its growth it will perpetuate its form and proportions. When the balance turns against the life-force, and from disease or age the assimilations cannot repair the lesions, the body must decay and the life-force become disembodied. As the life-force overruled and used the material forces, so in all cases of partial or total disembodiment, the material forces again take on their old unhindered working, and what was living body becomes again dead mechanical matter, and falls into the conditioned successions and changes of its old mechanical forces, and we have death and dissolution.

But as the death and dissolution of the individual occurs, provision must be made for the *generation* of successors and thereby the perpetuation of the species. From the ancestral stock there must be the setting off an instalment of the life force in a new germ, and which may begin its own process of assimilation and growth, and instead of stating or positing itself in its parental body, may separate itself wholly from that, and build up to maturity its own independent body, and thus the species propagate its successive generations. In order to such generation, the principle of *sex* is necessary. The life-force in the one sex must go over into some prepared receptacle of congenial nourishment from the other sex, and a double gender can alone procreate a new offspring. Nor is it consistent with the demands of reason that the original types of organic being should be marred and confounded by a promiscuous generation, and the necessities of nature will also second this demand of reason, for the receptacle afforded by one gender in a species cannot contain the appropriate nourishment for the living force imparted by the other gender in a different species, and hence opposite sexes in different species should not and cannot procreate, or if they do for once, the hybrid progeny must be barren. The individuals must have their separate gender, the species lives on in the generations of both its own sexes. Life must thus work on in cycles, and each species perpetuate itself in the perpetual propagation of new individuals. The propagated life-force, though beginning a new assimilation and incorporation of its own, will be still slowly exhausting the original energy, and thus at last the species must die out in the deterioration of its members, and new species must be put again into nature, to run their cycle conformed to the altered chemical combinations in the world of matter around them.

23. The Principle of Vegetable Life.—The life-force is in itself a spiritual activity which works according to wants, and therefore in reference to ends, and is first an assimilative force by virtue of its combining other forces in it, and then consequently an indrawing or inhaustive force by occasion of the vacuum which the vital combination secures between the living corpuscle and the material molecules about it, and the forcing in of these material molecules upon the point of vital action by their own inherent antagonist working, and thus affording a perpetual supply to the life-force for perpetual assimilations. Such combination of the life-force with the material forces thus brought into concretion becomes a germ, and has in it and with it all the elements and rudiments of the future mature organism. The life-force can make no possible manifesta-

tion of itself in its simple spiritual being, and must at once seek to combine itself with congenial material forces. Such combination in any way and to any degree of completeness will be a *germ*, and may go on to complete envelopment in matter, and then on to complete development in the adult organism, but that germ only which is constituted through the medium of the sexes, and which by its own growth separates itself from the parent stock, is properly a *seed*, and in which the species perpetuates itself by multiplying itself into innumerable successive individuals.

Now, the primitive created germ must have been this originated spiritual life-force from the Absolute Spirit and its combination immediately with the congenial material forces in the place where it was set, and in its simplest and earliest form of working we may trace it out from the already attained idea, and we shall have the principle of vegetation that must determine the laws in the facts of all coming experience. This primitive germ, and equally so with every matured sexually constituted seed, must have its want, and thus its end of acting. Disregarding here all specific wants, and which in their distinction only separate the different species, we seize only upon the generic wants, and thus attain the ends that must be common to all vegetable species, or the whole kingdom of plants and trees.

The first want, and thus the first end to be attained in the growing germ, is the perpetual and abundant supply of material forces that may be combined with it, and thus its first growth and development must be in the direction and out into the midst of such congenial material forces. These are found in the soil of the earth beneath and in the gaseous atmosphere above, and the very necessity of the germ must send its roots downward and its stock and branches upward. What grows downward will adapt itself to its destined ends in its constituted wants, and the root will have its distinguishing characteristics, and so what grows upward will in the same way conform to its end, and the branch, bud, and leaf, will have their peculiar characteristics. All radicles must thus differ from their stock and branches, for the wants of the whole plant make both necessary, and each a necessity in its own peculiar end for the grand general end of the whole, and hence each must be fitted for its own function, and the whole plant, root, stock, and leaf, becomes an organic existence, having unity and identity in every part, and the whole as truly a means for the end of each part as each part is a means for the grand end of the whole.

The material forces which are to be combined and assimilated, will be diffused through the soil beneath and the air above, hence the radicles will separate themselves on all sides through the soil, and the branches on all sides through the air, or go off successively, and more or less diffusively as the plant needs, from a tap-root below and a main trunk above. The life-force must thus grow out each way from its salient point, and assimilate and incorporate itself in the material plant, as all together in root, stock and branches, making one organic identity. This order of growth determines the working of the life-force perpetually to the surface. The supply of material to be used is both exterior and adjacent to the assimilative forces, and is immediately combined with them by perpetual accretions superficially, except as enclosed and defended by an exterior rind or bark, which is itself rather an exuvia than the product of a living process. The perpetual assimilations all along the course of the living process, must perpetually keep open continuous pores, through which the external forces of nature must be constantly supplying the pabulum for further accretions. The central part only of the trunk and roots will maintain itself in vitality and in solid consistency proportioned to the intensity of the fibre incorporated, but all new growth of the plant must be superficial. The ongoing of the assimilating process is still perpetually from the salient point between the rooting downwards and the branching upwards, and hence along the vacant pores all the way up the trunk and branches, the new matter for new assimilations is successively incorporated, and the terminations must be in a bud that perpetually turns itself outward in development, and maintains itself as a bud in constant self-reduplication. The bud must also perpetually leave its vitalized substance behind it, posited as a solid and extended branch, which henceforth has its concretions on the outside, as the matter for vital assimilation comes up in the pores that open on the points where the life-force is incessantly working.

Each bud is thus a complete germ, being a whole within itself, and each branch a complete plant, so that the whole vegetable organism, though in itself a unit, may multiply itself by slips, and grafts, and buds, and the one plant still remain in vigorous growth, while its detached portions may send out new roots in another soil, or have their pores kept open for new assimilations, by the life-force of another stock within which they may have been inserted. And though the root and branch differ from each other in their characteristics in their places, yet is this difference only in characteristic and not of kind, for by change of place there may be made a change of function, and the branch send off its roots, and the root send out its buds and leaves.

Such must be the principle of vegetative life in general, and yet it is manifest that many specific differences may abound, and both the demands of reason in securing general order and specific variety, and the adaptations in nature of fitting forces to be assimilated by only congenial spiritual activities, will make many differing species of plants to be a necessity. As the chemical development in matter goes on, the life-force will be interposed by the Creator; and the incorporation into a germ, and the development of the germ to the mature plant, and the multiplication of the plant in its many seeds, will all conform to the maturity and completeness of adaptation in the geological process. The earlier will doubtless be the least complicated, and while primitive species may be enormous both in number and size, yet the more complete and perfect types of plants must be reserved for the more mature and elaborate chemical and geological preparations. Many old species of plants may wholly die out, and have their being only in fossil preservations, before the more perfect types and specimens of vegetative life can be introduced, to incorporate themselves into sufficiently sublimated material elements.

24. The Principle of Animal Life.—The life-force elaborates the vegetable organism immediately from the chemically prepared elements in matter, and overrules and assimilates mere mechanical forces, and combines them by its own agency, as a higher form of vital chemistry, into a

new substance that may be called living body. But the most perfect and stupendous forms of vegetable life must be circumscribed within this sphere of superficial assimilation and inhaustive supply through the pores made void by the used-up mechanical elements. A perpetual indrawing and incorporating preserves the growth and health of the plant, but in this way the life can be that of the plant only, whether it be an annual or endure for centuries. By no cultivation and favoring conditions can the vegetable pass the boundaries of its proper jurisdiction, and leap within some higher kingdom.

But it is clear that reason cannot satisfy its own demands in making the vegetable an ultimate end. It is a higher form of existence than any merely mechanical force can constitute, and is thus an advance in nature; but nature might go on her ceaseless round of chemical changes and vegetable products, yet that one plant should live and grow only to produce another of its own kind in endless successions, would never equal what the Absolute Spirit must ask of itself for its own dignity and glory. The last individuals of the species would no nearer approach to any rational consummation than the first. Reason clearly sees the plant to be as truly and necessarily a means to an end as the working of mechanical forces, and with the mightiest and oldest oaks and cedars before it, the reason must still affirm, here is nothing that is end in itself, but the end is still beyond; what is this majestic, long-lived plant for? A higher living organism that may use the vegetable, and be its end, must be created. Nature wants and reason demands the higher animal kingdom.

Inasmuch as the animal is to use the vegetable, and its

assimilations and incorporations are to be of no mechanical and chemical elements only, but wholly of such elements vitally combined, it is clear that some peculiar characteristics must distinguish the working of the life-force in the animal from any thing we have found in the vegetable. The plants to be drawn in and assimilated, cannot be reached by any rootlets in the earth, nor any branches and leaves in the air. The first and great peculiarity of the animal organism must be, *that the vital force be transferred altogether from the surface to the inside.* The vegetable pabulum for its growth and preservation must be gathered and retained by the animal, and the work of assimilating and incorporating must go on within the organism, and hence there must be a stomach and intestinal canal, with its absorbing and imbibing mouths like the radicles of the plant turned inward, and through which the forces to be combined in the living flesh and bone of the animal body may be taken and brought directly to the points of the living and working forces. And not only the functions of the roots, but those of the leaves in the plant also must now become internal, for the whole work of oxygenation or imbibing that which must be used in a gaseous form, makes it necessary that provision be made for this through all the circuit where the inward assimilating process is to be carried on. The functions of respiration must be transferred from the outside leaves to the inside lungs, and while the chylification goes on in the stomach, and the prepared matter for combination in the living animal tissue is poured into the blood, and sent on through the whole arterial circulation, so the additional preparation of the oxygenating process must mingle in the same current, and transfer itself to the same internal laboratory. The life-force in the animal must be housed within, and build up over itself its own dwelling.

But still further, for the carrying on of these functions, and for the necessary locomotion in gathering and masticating its food, the animal life must possess the force of *muscular irritability.* The life-force has here new wants and must be made to work for new ends. The plant may so combine the antagonist forces in their polar relations, and so temper them in the surroundings of the diremptive forces, that external action upon it may induce a contraction and withdrawing from the fretting violence, or an expansion and approach to the genial influence, and thus something of the forms of muscular action may be given to the vegetable fibre. But in the animal, a much more unequivocal, positive, and extensive muscular force is demanded. The life-force must so arrange in combination the polar forces in the molecules which it assimilates in the fleshy body of the muscle, that by its own impulse it can on any occasion secure that every living corpuscle in the muscle shall react upon its own centre of antagonism, and thus the whole muscle contract upon itself at its centre by this reaction upon itself of every molecular force that constitutes it; and then it must so attach the muscle to the bony structure of the frame that it shall lift and move the member accordingly. Here is a much higher want, and hence a much more complicated adaptation to an end than any thing for which the vegetable kingdom can be made to legislate.

With such adaptations, it must immediately follow, that from the irritation of the contents themselves, as in

the case of the food in the stomach, the air in the lungs, or the blood in the heart, the contraction of the muscles shall spontaneously take place, and the want be satisfied. The forces used and combined by the life-force, are abundantly sufficient in their varieties and degrees, to secure any possible end of muscular contraction, and the Absolute reason has but to endow and adapt the life-force in its wants, and supply for it the congenial materials in the mechanical forces, and its own spontaneous working will determine the issue. The muscle will differ from any other bodily tissue, and be susceptible to magnetic and electric forces from its own polar arrangement of molecules, unlike any other forms of matter.

No animal muscle is self-active. It must receive the action that irritates it, and thus induces contraction in it, from some foreign force. Its irritation is not at all sensation, but a contraction from some outside invasion, and has nothing that yet awakes in any sense, much less any will. Sense and will may use muscular irritation and contraction, but the life-force also uses it in blind spontaneity, before it has itself been elevated to any of the prerogatives of a sentient and voluntary existence. Some forms of animal life will rise above the vegetable only in the transfer of the assimilative force from the surface to the inside, having merely an alimentary canal and a contractile capability through the whole body, while in other rising forms of animated existence we approach the completed types of organic structure in the possession of all the digestive and respiratory functions, the perfection of arterial circulation and free locomotion in the use of both the muscles of involuntary and voluntary action. From the crawling earth-worm to the solid tread of those animals whose organism includes the whole frame of spinal, vertebral, crural, and brachial attachments, we have all degrees of muscular fulness, and yet so far as muscular only, they have all but the same contractile irritability from an outside agency.

There must still be a higher form of vital action, or the animal want cannot be satisfied and its ends attained. Merely that the content itself in a muscular receptivity should irritate and induce contraction in the muscle will not be enough; the necessity must often occur that the animal itself secure the presence of the content, especially the aliment for the stomach, and thus the occasion for muscular motion be furnished before the irritating content comes, even in order that it may be attained, and to this end there must be *the nervous sensibility.*

The molecules in the muscle contract by occasion of some foreign action upon them; the life-force having so combined and attached them, that their contraction and motion are the result of direct mechanical forces already within them; but now the life-force must go further and combine for itself a material organ that can act, not merely when acted upon, but which can upon occasion being given act directly back upon itself, and such material organ will be a nerve of sensation. The vital force in the nerve makes the molecules of the nerve to act back upon its own agency, and in this reception of action from its own action, an impression on itself from its own movement, there comes a self-feeling or awakened sensation, which at once distinguishes the animal from all forms of vegetable life. The existence of a brain and connected system of nerves, gives to the life-force a self-centre from whence it can act and receive the action, and in this affecting or impressing itself in its own organism, there is self-reciprocity, and thus sentient life.

In a bodily organism so endowed, any want in the system awakens some appropriate nerve and makes itself to be known as an appetite, and every supply of the want also makes itself to be felt as animal enjoyment. This nervous sensation is itself an occasion of irritation to the muscle, and thus the life-force awakened to feeling in an appetite, at once impels the contracting muscle, and locomotion and self-supply succeeds. Not only will the appetite thus direct the muscle in locomotion, that it may bring the body to the place of its desired object, but the successive wants direct to all the successive motions, which work out their respective ends. In hunger, the food is not only attained, but all the muscles necessary for grasping, masticating, and swallowing, are set in motion by the feeling. The nervous system is the source of all the animal susceptibility.

When this reaction of the organism upon itself is slight, or when that organism is less intensely vitalized, there cannot be made a distinction of both the action and reagency, and the result must be a mere blind sensation only. The action will have more in it than mere muscular irritability, for it truly reacts upon its own organism, but yet as undiscriminated action and reaction there can be only an instinct, prompting and directing motion merely through the impulses of the sensation itself. But when this action and reaction is strong and in an intensely vitalized organ, and there is thus a capacity to distinctly apprehend the nervous reaction, and also that this reagency is itself an induced result, there must be given a conscious sensation, a recognized feeling, and in this the animal sentient life at once begins. Animal desire, and not mere instinct, must then govern the movement, and the locomotion is guided by the sense to the attainment of the objects which minister to the gratification of the desire.

In this there can as yet be given nothing beyond self-feeling; distinct sensation, but not self-knowledge. The agency and reagency of the organism upon itself is distinguished, and the feeling is a conscious desire; but the self is not recognized, and thus the animal cannot attain to self-consciousness.

When all these distinguishable alimentary, respiratory, muscular and nervous organisms have been superinduced upon the one life-force, and this permeates the whole structure which by its inhaustive and assimilating power it has built about itself, this structure of many parts must thus be made one organic whole, and the separate stomach and lungs, the irritable muscle and the sentient nerve, all have their membership in and identification with the one body. The vegetative force is turned inward in the stomach and lungs of the animal, and is thenceforth no longer merely vegetative. The superinduced force of irritability as an action and reaction of content and organ must be given for the uses and ends of this transferred inward vegetative process, that thereby there may be secured the presence of the constitutive matter, or pabulum, for the assimilating activity. Here will be found the lower forms of animal existence. The coral animalcula is scarcely other than a muscular organism vitalized, acting through its whole structure by inhaustion and irritation; assimilating and incorporating, and then absorbing and secreting, those excrementitious fabrications, that ultimately make the extended and solid islands of the tropical seas. Rising higher in animal existence, as the vegetable life is pushed further inward, we must have the nervous system for the use of the augmented life-force, and here we attain the comparatively elevated form of organic existence in a locomotive, masticating, sentient and percipient animal. This animal holds all the orders of inferior existence within itself and subject to its own uses. Neither the material nor the vegetative forces have been lost in the higher superinductions upon them, but we shall find them still unchanged except as made subservient to the higher activity. All become one in the one organized system, and these are successively built up each by their own forces, as creation advances and matures. The antagonist and diremptive forces make the material world, and the assimilative force makes the vital world, and the vital in the material builds up its own body superficially as the plant, and through the media of muscular and nervous instrumentalities, also builds up a body about itself from the inside as the animal. But highly as the existence has become elevated in the animal, and the idea of created being has been here advanced, still all is within nature, and bound in the conditioned and necessitated births and growths determined for them in their primal constitution.

25. The Principle of Human Life.—All creation is as yet means to ends, and the attainment is not yet made of an end that can be self-satisfactory and thus ultimate. The antagonist and diremptive forces work on and never finish;

the life-force builds up the plant, and pushes out perpetual buds in self-reduplication, and throws off its seed after its kind in the perpetuation of its species, but at no point does the plant turn back upon itself and come to any self-finding; and then the animal inwardly digests and respires, has locomotion and sensation, and a perpetual circulation of life and feeling about a centre, but in all this going out from and coming back to a centre, there is no capability to remain and retain itself at the centre. So soon as there is a self-finding there is also a self-losing, and thus only a successive self-feeling with no self-possession. The most intense animal sensation is perpetually transitional, and never comes to any abiding self-consciousness. All is thus nature; conditioned succession; determined but interminable births and deaths; and as yet nowhere the capability to rest in any consummation.

With this reason cannot be satisfied. The Absolute Spirit cannot rejoice in his own work except as it is made at last complete in itself, and possessing that which has an intrinsic excellency that may properly use and exhaust for itself all this universe of created means, and be an end in which they are swallowed up. Without such a crown on nature, her last birth and growth is wholly meaningless, and there has been nothing to work *from*, and nothing to work *for*, and therefore nothing worthy of the Great Architect to work *out*.

Superinduced upon this animal life, there must be the force of reason, which can read principles and law in itself and control all animal feeling by them, and hold all of nature that is in him freely and joyfully subject to them. Such a union of the animal and the rational will be *the hu-*

man; not thing, but person; in nature, and yet supernatural. While he can use all nature's means for his ends, he can also know and commune with the God who reveals himself in nature, as partaking himself of his likeness. God may not only express himself in him as in all his works, but may reveal himself to him in ways which none of his created works can express, and bring him thus intelligently and eternally in adoring communion.

Nature in the individual has ends for which it works in the completion and preservation of the individual; and even ends beyond the individual in the perpetuation and melioration of the kind, and which reason would itself dictate, and thus nature and reason are in these ends at one, and the whole humanity may go out spontaneously to gain them. But the ends of the reason reach far beyond the point to which the animal may follow in individual happiness and the *well-feeling* of the race, even to the moral and immortal *well-being* in constant self-approbation and divine approval.

Should the rational bow in bondage to nature as it can, and take the end of the animal as its chief good, it then becomes merely a servant to nature; a bond-slave to the flesh; but cannot thus become nature, and put off its spiritual prerogatives and personal responsibilities. The conscious obligation must still press, that this bondage cease, and the body be at once brought in subjection, and nature put to serve the spirit, and not that the spirit may put itself for one hour to yield its end to the lusts of the flesh.

This superinduction of the rational must perfect and consummate the animal. The intellectual life through the sense and the understanding must be thereby lifted above any illumination which mere animality can reach. The bodily organization must also find in this its perfection. All along in the plant, the most perfect of its kind is when the interpenetrating life is the most intense and unhindered. The animal body is most complete when all the forces of vital assimilation, muscular irritation, and nervous sensation, are the most freely active. The life-force is the most energetic in the germ that contains the rudiments of the most complicated organization, and the superinduction of higher forces must ever elevate the life in the corporeity it inhabits. We must, therefore, find all forms in nature below ripening upwards toward man. In him must be the consummation of all corporeal organism. The archetype after which nature has been working comes out at last in the human form, and all lower bodies must possess their rudiments in nascent progression till they culminate in the erect stature and expressive countenance of man. He holds dominion over all the material, vegetable, and animal creation.

CHAPTER III.

THE NECESSARY LAWS IN THE ACTUAL FACTS OF THE UNIVERSE.

The universe in its eternal principles gives the creation in Idea, and in this we know what is possible; but an insight of the universe in its principles does not warrant the affirmation that what is so clearly possible to be, actually is. A universe so may be; yea, if a universe of working central forces be brought into existence, so it must be; but that the universe shall so be in actual fact, there is demanded the exertion of creative Omnipotence. The created facts being given, the reason may in them detect the laws by which they are governed, and when the insight of reason also determines that these very laws in the facts are such as the eternal principles made necessary, we have then a true and valid science of the universe, and may safely call the result of our work a Rational Cosmology. This attainment of the laws in the facts, and their determination as necessary from the foregoing principles, is the business of this chapter, and in the accomplishment of which our whole work is completed.

It should be distinctly seen that the Creator himself cannot at all be subject to science according to such a process. He is the absolute Author of all that has been so made according to eternal principles, and as unmade himself, we are not to search back of him for the principles which determine his possible being, and then search within him for the laws in the facts which such prior principles made to be necessary. Enough that in creation we find unequivocal certainty of facts that originated out of nature, and not as productions from something already in nature, and that the laws of all nature confirm the origination of the primal and constitutive forces of the universe to have been from a Rational Spirit, and then we have the demonstration both *that* a personal God is, and in many positive particulars *what* this God is; but we do not need to attempt the circumscription of him by a science, that in comprehending the facts in their principles shall enable us to say *how* God is. God is object for the insight of the reason, and not at all object for the judgments of the connecting understanding. The doom is on this latter faculty that its deepest substance must still stand in some lower substratum, and its highest cause must be the product of some previous causation; and from the law of its being and working this faculty must be shut up within nature, and can by no possibility take the leap outwards to a supernatural and absolute Deity. That a God is, and what a God is, is enough for rational faith and practical religion; but how God is will never be comprehensible within the deductions of a discursive philosophy.

The universe is alone the province and object for science. That is wholly fact; a thing made; and in all its particular facts there are laws which are necessary from the eternal principles which determined them; and thus here only need we look for laws in the facts in what remains of our work, just as here only have we attained principles prior to the facts in that part of our work which has been already accomplished.

Some of the first sections of the last chapter attain to principles of so general a bearing, that we do not need to refer to them in this chapter, and attempt to find any particular laws in the facts that have been determined by them. They have their application more or less to all facts, and give law to all facts, and will thus be sufficiently recognized in the recognition of the universe itself. Such is the case with the general principle of the space-filling antagonist force; the determination of space and time; the impression of matter upon the senses; that creation must be a nature; and the universal principles of motion. It is not until we come to the 8th section, that we need to take up the principles, item by item, and see how exactly they have determined the laws which we may find everywhere in the corresponding facts of the universal cosmos, as the careful observation of philosophical experiment gives them to us.

1. The Law of Material Sphericity.—The principle given in this section (Ch. II. 8) is, that all free matter, fluid or gaseous, must tend towards a globular form. The space-filling forces, generated from a constant antagonist action, must take on this globular form, from the necessary working of the countervailing activities at the centre. The immutable laws of motion secure that, as new forces are generated by each activity pushing the other back upon itself, so these generated forces must every way expand from the generating point, and successively, layer after

layer, ensphere themselves about it. The Absolute Creator could make matter take on other forms, for he could combine single activities in other numbers and directions of working, and thus induce other forms of static equilibrium and thereby give other permanent shapes to matter; but if a direct antagonism be taken as the generating force in creation, then the free matter, so fast as created, must range itself equably about this generating source and become a sphere. The origination of space-filling forces in that manner must fill space in a growing sphere. This matter, or space-filling force, must preserve its direct antagonisms so long as its existence remains. Other forces may be superinduced and widely vary its consistency and fluidity, but the primitive ethereal matter must ever tend towards a globular arrangement of parts, and wherever matter shall be left free in its movements before its own constituent forces, it must ever be found with this inherent law of sphericity. The universal working of the component molecules, in their several antagonisms, must give equally balanced centrifugal and centripetal tendencies to every separate position in the space filled. And now this principle in idea is everywhere the law in fact.

We have nothing to do with solid matter in its rigid state, inasmuch as its component molecules are not free to take their positions according to their concentric tendencies. What the tendency in all solids is, and what the result would be if free to conform to this tendency, is manifest from the fact that every detached portion of matter has its own central balancing point. Angular as may be its surface, there is always a plane that equally divides the intensity of its forces, always an axis that equalizes its polarity, and always a centre that may sustain all its outlying portions. Were the mass a free fluid, this fact of self-balanced parts evinces that this plane would at once become the division between two hemispheres, this axis a bisection of that circular plane, and this centre the point of equal radiation every way to the circumference.

But take any fluid matter, and give the occasion for its own free movement within itself, and its law of sphericity will universally appear. The rain will fall in spherical drops until flattened or spattered by meeting a resisting surface. The dew-drop will stand ensphered on the leaf, and only flattened on the side of its support to just the equilibrium of inherent consistency and specific gravity. The drop of quicksilver, though of much greater specific gravity than water, yet from its excess of inherent consistency will not permit that its weight should flatten it on its support so much, but it stands up in relief from its basis almost as perfect a globe as if freely suspended by its centre. The central antagonism works as the same law in the water and the mercury, and perfectly enspheres both, but the less consistency of parts in the water gives occasion to the weight to make greater interference with the law than in the mercury. All fluids are flattened and not spherical, only because their masses make the aggregate gravity to overwork any common central antagonism, and they thus spread their parts on the beds that sustain them. The proper radius would always secure the sphericity of the fluid.

Vapors, in their most volatile state, have the same law of sphericity as fluids. If the vapor be denser than the atmosphere that surrounds it, then will its parts come under the law, and ensphere themselves about a central antagonism, subject to all the disturbing forces in the agitation of the surrounding atmosphere, and the flattening from the specific gravity of the mass as it compresses itself on the substance that supports it. If the vapor be lighter than the atmosphere in which it floats, then must it be subject to the forces in the atmosphere, and the excess of force without must overbear its antagonist working within, and it can only take the shapes the outward forces impose upon it. But if that volatile vapor alone occupied its own space, it would as truly evince its law of sphericity as does the atmosphere itself, or as does the primitive ether to which the atmosphere is only a much denser vapor. The molecules of vapor no further fly apart by what is termed repulsion, than is demanded from the force of the central working, which must expel from a given centre to the point which exactly balances the reaction of that which urges towards the centre. The vapor is only a lighter fluid, and all fluids here have the same determined law. The shot, which has fallen two hundred feet and cooled to a solid in its descent, was rounded to a ball in the first portion of its passage by this inherent law of its sphericity from its own constitutive forces. Not at all because pressed from the outside has this form been induced, for this outside pressure has not been on all sides equal; the shot would have been the same perfect ball if it had fallen through a vacuum, and provided it could thus have parted with its heat, it would also have been the same solid. The law of ensphering is from within, even from the working of its own constitutive forces, and not that any other matter has been doing the forming work.

There may be globes which have been chipped off from the outside when already solid, or that may have been moulded by an outside pressure, and such will ordinarily carry with them the traces of the external violence; but where a solid is found as a globe in a natural state, it is safe to conclude that when formed it was a fluid, and allowed its central force to permeate its whole structure, and arrange its particles according to the inner law of sphericity. So the worlds of our solar system were formed, and though now mostly they have become solid bodies, yet the fact of their globular forms carries conclusive evidence that they took on their shapes when fused, and free to move according to the tendencies of their central forces.

It is more surprising to find the facts of what is called *capillary attraction* unexpectedly leaping within the control of this law of fluid sphericity. If a glass tube of small diameter be plunged at one end perpendicularly in water, the water in the tube will rise considerably above the surface of the water on the outside. To say that the inner surface of the tube *attracts* the fluid, is only to cover the mystery under a word, and explains nothing. The atmospheric pressure will make the water in the tube rise to the height of the outer surface, but such force can do no more. The superficial stratum within the tube is acted upon, through all its molecules, by the constitutive antagonisms which tend to ensphere themselves about the central force. The inner surface of the tube is as a surrounding wall, restraining this ensphering action, and thus necessarily forcing the fluid further up, and making what would have been a sphere to be compressed within a smaller diameter into a cylinder. The fluid must rise, until what from the central force would have been a sphere must now find its balance in the longer axis of the cylinder, and at that point the water must stand. Hence the smaller the tube the higher the water must rise, for the central pressure that would have ensphered, must find its balance in a longer axis of the smaller cylinder.

If two plates of glass be joined in an acute angle at one edge, and open at the other, and these so plunged perpendicularly in water that it may rise between the plates, we shall have the water within rising above the surface of that without, and at an elevation proportioned to the nearness of the plates to each other. At the angle it will be of the greatest elevation, and regularly diminish in height to the outer extremity, making the curve to be that of a parabola. And here we have precisely the same law, for the central force which would ensphere the included water is now resisted by the glass walls, and the water, which would have gone out all ways from such force, must now rise between the glass plates, and the higher in proportion to their contiguity. These plates are as the inner surface of the tube, and the antagonist working within gives the same law of rising in height, proportioned to the diameter of the space within which the fluid is pressed; and inasmuch as the pressure does not surround the fluid as a tube, but only partially incloses it as by two sides of a triangle, so the height can at no place be but to one-half the amount between the plates that the same diameters would have given in a complete cylinder. The facts and the determination of the principle perfectly correspond.

If the tube be a long slender cone, and a small portion

of water be introduced at the base, then what would by the central force have been ensphered in a drop, must now take on the form to which the sides of the tube compress it; and as this is perpetually diminishing upwards, the central working must force the fluid in that direction, and make the portion of water inserted at the bottom move progressively to the top of the tube, as the facts disclose.

If the fluid be in a large glass vessel, so that only the contiguous surface resists the ensphering force, then we shall have the common fact of the surface of the fluid made somewhat elevated in contact with the vessel, and curving off concavely at the departure from it. If this fluid be of so great consistency that the ensphering force cannot drive it up against the side of the glass vessel, then instead of rising against, it must tend to ensphere itself back from the vessel, and we shall have the fact as in mercury. When there is the absorbing of water as in a sponge, or a loose string or cloth, the contiguous threads or filaments are as the sides of the tube, and the ensphering force pushes up the fluid between them, and as there are constant cross filaments so the fluid is perpetually supported, and giving occasion for continual progress through the absorbent.

2. The Law of Gravity.—The force engendered by the primal antagonism, must not only ensphere all the successive points of force engendered, but must so ensphere them that each point out from the centre must react back upon the centre in exact static equilibrium. The central repulsion is every way equal through the sphere, and directly as the intensity of the forces, which is the quantity

of matter, and inversely as the cubes of the radii, which is the distance. The reaction, which is the attraction of any point in the sphere, is in the direct line of the radius towards the centre from that point, and as the intensity or quantity of matter directly, and as the square of the radius or distance inversely. Each point in the sphere is thus determined in the density of its matter, or the intensity of the force which fills it, by the intensity of the central antagonism, and this also determines the magnitude of the sphere. This reaction, or attraction, of each point towards the centre is the result of the compounding of the sphere-forming agency, which from the centre so pushes out every point of force in reciprocal action with every other point, that each is both pressed, and in return presses every way. All the particles of the sphere, therefore, both repel and attract every other particle, and in the above ratios for the repulsion and attraction.

This must be so in all parts relatively to the universal sphere, in all parts of each particular globe relatively to its own sphere, and of all particular spheres relatively to each other. All matter in every way relatively to all other matter must, therefore, gravitate toward all other matter, directly as the quantity and inversely as the square of the distance.

Now, that the law in the facts accords with this necessary principle in the reason's idea, needs no otherwise to be noticed here than by a general reference to physical science. Long, exact, and extended observation and experiment have found the facts to be thus, and with not an apparent exception, and all inductive philosophy rests upon it. Whether in our world or others, so far as facts can be gained, the facts come within the control of this law, and this has become so universally admitted that any detail would be not only superfluous but intolerable. The only thing important is to correct the error and remove the absurdity of the common notion of gravity, as itself a fact assumed from this apparent concurrence of all facts. All observed facts of matter have resolved themselves into or rather bound themselves by, and comprehended themselves within, this higher fact, that all matter tends to all other matter in the above gravitating ratios.

This fact has been assumed as an ultimate fact from all observed facts, but no way opens to the attainment of any point of observation that may make this ultimate fact a matter of experience. It can, therefore, never be experimentally verified nor expounded. The conception of it can be the product of the imagination alone, and that it may not be mere fancy, but a logical product of the discursive understanding, the conception is thought out from the facts observed in the following process. If there were no other forces acting upon matter than attraction, or the tendency to come together, then it follows that all matter must have long since become consolidated. Such, however, is not the observed fact, but worlds stand apart from each other with no observed matter to separate them. They are found in distinguished cases to revolve one about another, and many substances in our world repel other substances; there is then another force the opposite of attraction, and by compounding the movements induced by such forces, the imagination forms the conception how worlds may be kept separate, and yet in general connection. These opposite forces are then deemed to be so arranged that the compounded motion shall be in orbits about a common centre. All matter may thus continually gravitate toward all other matter, and yet instead of coming together, one portion of matter may move around other portions. The conception of gravity, becomes thus a centripetal force, and this balanced by a wholly independent and opposite centrifugal force. Whence these forces come, is as inexplicable as whence the matter comes; but the conception is of matter, that it is itself inert, and the centripetal force is added to make matter in some cases come together in bodies, and the centrifugal force is separately added in other cases to keep the bodies apart, and the two forces are in other cases to be compounded into revolving movements. These are the imagined facts, as thought out logically from the discovered facts, and thus far only can the discursive judgment, or the logical understanding, frame its imaginings to explain its universal fact of gravity.

But precisely here, as in all processes of the logical understanding, the explanation can possibly do nothing but interpose the sophism of a *petitio principii;* and when the sophism is exposed, there is nothing left but the alternatives of direct skepticism on one side, or the running to an eternal series on the other. The bald skepticism, the blank doubt whether philosophy knows any thing, is usually too humiliating and unsatisfying, and the common resort is to run up a series for some successive links, and then leave the mind to delude itself by a sophism that there is somewhere up in the indefinite obscurity an absolute stand-point, which holds all fast and makes all plain. So we have our central earth and revolving moon; and then our central sun and revolving planets with all their moons; and we can stand on our earth, and judge the moon to have a stable position from which its gravity does not need to endanger that it shall fall off; and then we can take our stand on the sun, and the earth that holds the moon is also stable; and then when we ask for the sun's hold-point, we at once send it revolving around a higher centre, with a proportionate augmentation of repulsive and attractive forces; and then this around some higher and greater; and finally lose ourselves in looking at, without attempting the actual passage through, this indefinite regressus. Or perhaps some one brings in the Deity as a grand gravitating centre, and says of the last great world that it revolves about the throne of God; just as the Stoics said of their empyrean vortices, that all were whirled from the great central pyramid, which was the watch-tower of Jupiter. We are seeking a substantial static for all substances; a central point for all gravitating forces; and apparently unconscious that the very attempt involves an absurdity, we rest at last in a worse because profane absurdity, and of which profanity we seem equally unconscious, that the spiritual Jehovah can stand in the place and be degraded to the instrumentalities of material gravitation. Thus it is that the logical understanding cannot come to any science, because it cannot find and know any ultimate. Its last and highest is always yet a fact that can have no explanation, because it cannot examine itself in the light of an eternal principle. Go as far as we may with the empirical conception of gravity, we must at last have a centre that will fall down, and a circumference that will fall off. As the only corrective to this, the law of gravity must be an insight of the reason, and seen as the one central working that generates both repulsion and attraction in the antagonism that is constitutive of matter itself. The centre and the circumference then alike hold each other.

3. The Law of Falling Bodies, and Material Pressure.—The principle of sphericity from the central force of antagonism determines the laws that must regulate falling bodies, pendulum oscillations, and fluid pressure, and which is but the application of the principle of gravity to the particular cases. The necessary ratio of increasing velocity in the falling body, both in its free descent and down an inclined plane, has already been determined in the application of the principle, and the facts in all actual experiments exactly accord. We have here no occasion for particularizing the facts, inasmuch as one invariable law prevails, only making the allowance for the density and thus the hinderance of the medium through which the body falls. The stroke of the falling pendulum, and the rise on the other side of the perpendicular from the impulse, and the ratios of velocity and extremes of oscillation in pendulum rods of unequal lengths, have their determination from the same principle, and all experience finds the accordant law in the facts. So also with the pressure of fluids, both in one vessel of either perpendicular or inclined sides, and in any number of vessels or branching compartments that have their free connections, the principle of sphericity from the central force determines the pressure of every fluid globule, and thus also the rise and relative surfaces in each compartment, and both principle and fact in experience perpetually coincide.

The only important thing to note in all these cases is,

that the laws are no generalizations from many conspiring facts, but the facts are seen to have such laws from the necessary determinations of the principles in which they are all expounded.

4. The Laws of Magnetism.—We shall find here many different applications of the principle in varied circumstances, and thus as many separate laws as there are distinct circumstantial facts. The principles of magnetism are given in that part of the force of sphericity which works from the equatorial plane on each side out to the poles. The central working must force the antagonisms at the equator in each spherical layer to flow out in contiguous meridional lines quite to the polar points, and thus every molecular force in every meridian becomes statically balanced in this equilibration of equatorial and polar pressure. Each molecular force must also have the line of its antagonism, or particular polar direction towards the centre, determined by its place in the meridional line, viz., parallel with the central antagonism in the equator; perpendicular to this central antagonism, or the main axis, at 45° from the equator, or midway to the pole; and turned quite round toward the centre, in the line of the main axis, at the pole. In this meridional direction of the force from the equator is the principle of magnetic polarity; in this changing molecular direction towards the centre is the principle of magnetic dip; and the compounding of the forces of opposite polarity and dip is the principle of magnetic attraction and repulsion; each of which is plain to a clear rational insight.

It is further carefully to be noted, that the molecular

forces, as mere antagonisms, become chemically modified in their combinations with the diremptive forces or heat, and that as such new molecular substances are combined, they also become more or less separated from each other and insulated in the permeating and dissolving action of the diremptive or heat-forces. Every molecule of matter must have, however, its particular antagonist or polar force, while it may be in contact with some molecules on some sides, and separated by the infused heat from other adjacent molecules on other sides, or may be wholly insulated by the heat-force from all adjacent molecules on all sides. With this full apprehension of magnetic principles in polarity, dip, and reciprocal dynamic influences; and also the more or less separation of each molecule from others by heat; we are prepared to take the facts and their laws, as given in actual experiment, and see how these laws in the facts are necessarily determined in the eternal principles.

Some bodies are found by experiment, when brought within the influence between the two poles of a powerful horse-shoe magnet, to arrange themselves in the same plane with the axis of the horse-shoe magnet, and such are called *magnetics;* other bodies so placed are found to arrange themselves at right angles to this plane of the magnetic axis, and such are called *dia-magnetics.* It is also found that some bodies manifest no susceptibility to the magnetic influence when placed as above, and such are said to be *indifferent.* Careful experiment further discloses that the magnetics are of the most chemically compact substances, or such as have the greatest number of chemical elements in the same volume of the substantial combination, and the dia-magnetics are the least chemically compact, with few exceptions. The magnetics may be mentioned in the following order of diminishing energy: iron, nickel, cobalt, manganese, chromium, cerium, titanium, paladium, crown-glass, platinum, osmium, and oxygen. All other metals, and all other matter, except atmospheric air, and such bodies as are carefully compounded to an indifferent state, are dia-magnetics. The three first magnetics have 230 chemical elements, and the others 170, in the same volume of substantial composition as will contain from 150 to 74 elements in the dia-magnetics. The marked exceptions are of the chemically compact substances, copper and zinc, which are dia-magnetics. Among the most readily dia-magnetic may be put, in their increasing order, ether, alcohol, water, mercury, flint-glass, tin, antimony, animal flesh, phosphorus, bismuth. Copper, which is chemically compact and yet dia-magnetic, becomes magnetic when combined sufficiently with oxygen, and may be so carefully proportioned as to be indifferent. A tube of atmospheric air, and an exhausted or void tube, are both also indifferent.

The principle determines and expounds the facts. The compact magnetics have their molecules so bound by their central antagonism, and so little dissolved in the diremptive action, that they are as a unit in their magnetic working, and turn as one body under the magnetic force; while the dia-magnetics have their molecules so insulated in the diremptive infusion, that each molecule separately obeys the magnetic control, and those on one side of the body are swayed by one pole of the magnet, and those on the other side by the opposite pole, and thus the substance as a whole must turn and rest at right angles to the magnetic axis. If the substance has such combination in itself as exactly to hold the molecules together in such proportion as to neutralize their aggregate magnetic action, the body must become *astatic*, or indifferent. A body may be chemically compact, as copper or zinc, and yet the compactness may admit of the insulation and free movement of the joint molecules in such separate pairs or parcels as shall put half the aggregate on one side and half on the other, and make the compact body still dia-magnetic; and then a combination with other magnetic molecules, as of the copper with oxygen, may bind the pairs or parcels as one, and either energize as a magnetic or just neutralize as indifferent.

Again, if steel-filings be evenly sprinkled on a pasteboard plane, and held horizontally over the poles of the horse-shoe magnet, they will at once arrange themselves in circular lines of collection, having a common diameter midway between the poles and at right angles to the magnetic axis, and passing each way around over the poles, and as the circles enlarge outward, their tendency is more and more to extend the circuit beyond the poles, and then to turn in each way upon the pole as if to combine and pass down in the magnetic axis to the centre. So, manifestly, the principle of magnetic polarity determines the facts must be. The steel-filings on the pasteboard are the index of the forces in a magnetic sphere, as if the sphere had been bisected from pole to pole through its axis. The magnetic meridians of each concentric layer, on opposite sides through the globe, are here laid open, and we can trace the path of the central working outward, in the cropping out edges of the successive layers, to the circumference. The opposite polarities, the slight gatherings of equatorial, and the rapidly accumulating collections of both the polar attractions, and the turning dip over beyond the poles and passing down the axis on each side to the centre, are all perpetual laws in every experiment, as the principle determines the facts must be.

The magnetic force is always manifestly in and with the magnetic body, and can never be separated and retained as if one magnet had been exhausted by what another may have received; and if a magnet be divided, or broken in many fragments, each portion is still at once a whole with its own polarity, dip, and attractive and repulsive influences. The very force of magnetism is its space-filling, matter-constituting force; not a force from some inscrutable source moving amid dead atoms and registering itself in their arrangement, but a force which constitutes and is the molecular matter, and thus of course the disposer of it in the necessary law of its own movement; and therefore always inherent in all matter as truly as gravity itself, and in the circumstances as really obeying its law in the diamagnetic, and in the indifferent body, as in the magnetic. And as every divided and broken portion of a body at once thereby attains its own centre and makes itself to be a new gravitating whole, so, necessarily in the same way, must each fragment be a new magnetic whole, in the instant outgoing of its equatorial and polar activities. The law is but that which the prior principle has determined.

A substance that is a magnetic is often found to be in such a state as to give no indications of polarity, dip, or attractive and repulsive power, and such substances are said to be *quiescent*, or in their *natural* state. But when such substances are placed in proper positions relatively to an active magnet, they become instantly or more gradually actively magnetic themselves, and such are said to become actively magnetic by *induction.* If such induced activity be slow and apparently with much difficulty awakened, and when so induced, if it remain in constant activity for a long period, the substance so resisting and retaining is said to have *coercive* force, and when the induction is instantaneous on the presentation of the active magnet, and as instantly ceases when removed, the substance is said to have *no coercive* force.

As illustrations of the above, there may be given the following facts. If a bar of steel, in a quiescent or natural state, be in certain specified methods subjected to the influence of a powerfully active magnet, it will itself gradually become actively magnetic by induction, and will retain the constant activity for a long period, thus manifesting a high degree of coercive force. But if a bar of soft iron be so subjected to an active magnet, it will itself be instantly active by induction, and on withdrawing it from the permanently active magnet, it will as instantaneously become quiescent, thus showing that it has no coercive force.

The eternal principle in the reason, which determines all magnetic force, at once reveals and expounds these necessary laws of all magnetic induction. When the molecular forces in the magnetic substance are so dissolved by the permeating diremptive action, or force of heat, as to admit of their being turned in their particular polar directions in all ways promiscuously relatively to each other, then is the substance magnetically quiescent; but on the presentation of a powerfully active magnet, every molecule in the quiescent mass is made to feel the force of this magnet in its attractions and repulsions and to assume their relative polar directions accordingly, and thus the whole bar has its molecules so arranged that as a unit it acts magnetically. If the molecules were so slightly dissolved and insulated by the combined heat as to admit of their changes of position relatively, only with great difficulty, then was there much coercive force, as in the hard steel bar, and they will retain their magnetically active position with proportionate tenacity; but if the molecules were so dissolved and separated as easily to turn into their symmetrical polar positions, as in the soft iron bar, then also will they lose their magnetically active position as readily. When the steel bar is under the inductive influence, if it be repeatedly struck, and its molecules suddenly agitated by the blows, its magnetically active process is very much hastened. So also, if the soft iron bar be hammered and thereby considerably condensed, while under the inductive influence, the molecules are thus hindered in their change of polar positions, and there is at once given a proportionally coercive force and a retention of magnetic activity. Heat also, when strongly applied to an active steel magnet, so far dissolves and loosens the molecules and destroys the coercive force.

In this induction of magnetic activity, the respective poles arrange themselves invariably in the order of an opposite pole in the induced magnet nearest to whichever pole it may be in the inducing magnet. Thus if the boreal pole of the active magnet be applied, the end of the bar nearest to this in the induced magnet will be an austral pole, and *vice versa.* If the active pole be put in the middle of the bar, the induced magnet will have one pole common to the two halves of the bar and of the opposite kind to the applied pole, while the two ends of the bar will be alike and of the same kind as the applied pole, making the induced bar, indeed, to become two magnets. And if the active pole be applied to the centre of a circular plate of soft iron, the central portion will be an induced polarity opposite to that of the applied pole, and the whole circumference of the soft iron plate will have a polarity like the applied pole; the whole plate, indeed, becoming so many distinct magnets in every radius, the central ends being all of one polarity opposite in kind to the applied pole, and the outer ends being all of one polarity and the same in kind as the applied pole.

The principle of magnetic attraction and repulsion also determines this law of inductive polarity, for the applied pole with its specific direction and dip must repel similar polarities and put them the furthest from itself, and attract opposite polarities and place them the nearest to itself, and thus make the neutral point, or centre of the magnetic axis, midway between the two induced poles. And this same principle of magnetic attraction and repulsion determines the law in another very remarkable series of polarities given in peculiar cases. If the quiescent bar have all along its length inequalities of density, then will the induced magnetic action form itself into successive centres in these denser parts, and with their boreal and austral poles the nearest to the applied pole of an opposite kind, and the furthest from the applied pole of the same kind; thus making an induced magnet to have what is called *consecutive polarity.*

The magnetic axis is often other than the geometric axis of the same body, and this fact is also explicable by the same principle. If the molecules were all equally free, and all equally energetic in polar force, then must the magnetic and the gravitating centres be the same, and the arrangement of molecules from the centre be the same, and thus in each case the magnetic and gravitating axes must be the same; but if some of the portions of the magnetic body have facilities or hinderances in the changes of molecular polarities in different degrees on opposite sides or ends, and these do not correspond with the chemical combinations in the same portions as to their density and gravity, then must the magnetic and geometric centres and axes differ, for the inductive force cannot then follow and conform itself to the force of gravity. Thus the aggregate forces in the chemical combinations and the molecular polarities, the one as gravity and the other as magnetism, may have considerable disparity, and thus disagreement in the position of their respective axes, and yet no modification of the chemical combinations and molecular polarities will be likely to occur that will very far sunder the one axis from the other, though constant changes may make perpetual variations in polar directions.

The laws of terrestrial magnetism are determined from the same principles as the laws of magnetism in any detached terrestrial substances. Indeed, the earth is to be considered as the great fountain of magnetic force, from whence all separate smaller magnets on the earth are induced. As the universal sphere of the primitive ether must be magnetic, so all globes, the base of whose matter is this primitive ether, and which have been formed in the free action of central forces, must be also magnetic. So the fact is, and so all the laws in the facts of terrestrial magnetism are found to be. The earth is a magnet; of opposite poles; its poles repel similar and attract opposite poles; its forces of attraction and repulsion are in the ratio inversely as the squares of the distance, and the attractions and repulsions, neutralized at the equator, are augmented gradually each way to the poles; and the magnetic dip, determining these attractions and repulsions, is conformably of opposite poles on opposite sides of the equator, and from a tangent to the globe at the equator, successively inclined towards and athwart the circumference as it approaches the poles, till at the poles it becomes a direct line in the axis toward the centre.

So also the earth is an inducing magnet, and iron and steel bars become magnetic precisely from its induction as from other smaller magnets. It has its non-conforming gravitating or rotating and magnetic poles and axes, and an equator as magnetic cutting twice across its geometric equator. While the geometric lines, polar, equatorial and meridional, are all regular and conformable, the magnetic lines are all slightly irregular and unconformable each to each. The axis is not with regular polar extremities, the equator is not in a uniform plane, and the isoclinic and iso-dynamic lines are not in any straight direction. The whole action of the terrestrial magnetic force evinces that it is following a law determined by the modified molecular polarities somewhat differently from the gravitating and geometrically sphere-forming forces. Hence its gradual polar and other changes, and which have not been, and from their heterogenous sources can never be brought into any conforming cycles with the geological movements. Isothermic and iso-dynamic lines may be nearly conformable, for the equilibrations of light and heat, which modify magnetic force, should there leave the magnetic force in its most unhindered and equal action.

5. The Law of Electricity.—Inasmuch as the earth is a magnet, so must it also be the great source of electricity for the experience of those who dwell upon it. The earth may have its interruptions, in its regular magnetic arrangements of molecular polarity, quite deep down in some of the meridians of its inner spherical layers, but if these should occur they would be out of the reach of our observation, and perhaps the superincumbent pressure may nearly or altogether prevent such internal derangement. Our experience of electrical phenomena must be of such only as are superficial in respect to the earth, and indeed in all electrical appearances in any matter connected with the earth, the phenomena will be on the surface of the material bodies, for the outer layers become an occasion for the ready restoration of any deranged polarity that may occur within the interior structure. So also the earth readily receives and neutralizes all electric tension that is positive in any portion of the atmosphere within about four feet of its surface, and below this line, atmospherical positive electricity is not found.

But in the surface of the earth, and the material bodies upon it, and in the atmosphere and its vapors, we have all the occasions for actual polar derangement and magnetic interruption, that we were obliged to take only in anticipation when we considered the eternal principle in the

Idea itself of electricity. Every terrestrial molecule may be separated from others by the permeating action of the diremptive force, and such isolation by heat gives at once occasion for the molecule to change the true meridional polar direction, and so far to interrupt the static equilibrium, and thereby necessarily induce an electrical tension in the contiguous molecules that are in their true polar direction. We have here, therefore, all the desired opportunity for actually applying to the facts the determinations of our previously attained principles.

Mechanical attrition, thermal expansion, and chemical changes, make derangements of corpuscular polarity, and in such interruptions of regular polar action, there comes at once the tension in the neighboring molecules which gives rise to all the phenomena of electricity. As these molecules are within the body of the earth at its surface, or in detached bodies upon the earth, or as in the atmosphere itself and the matter surrounded by it, so will the peculiarities of the electrical phenomena be varied; and our present business is to see, that these variations in the facts have their necessary laws determined by the eternal principles we have beforehand apprehended.

When this electric tension is excited, if the tension be restrained and cannot pass on and restore the interruption, the electricity is said to be in a *static* condition, and when it moves on in forcing out this derangement, the electricity is *dynamic*. One substance excited to an electric tension must press upon the molecules of a contiguous substance, and if they are in such a state of free polarity as to be modified by the influence, and become themselves electrically intense, such secondary bodies are said to become electric by *induction.* All bodies are somewhat susceptible to this electric tension by induction, but some very sluggishly and slightly, and others very readily and strongly; the former are called *di-electrics* or non-conductors, the latter are electrics, but usually denominated *conductors.* Di-electrics, when surrounding bodies under tension, or interposed between them and other substances which they might influence, thus cutting off their progress and keeping their electricity static, are said to be *insulators*, and the static electrical body is said to be *insulated.* Some bodies when excited always give a positive electric tension, and others always a negative, and some bodies vary according to the manner of the excitement; usually positive when excited by the harder or more polished substance, and negative when excited by a softer or rougher substance, but no examination of the one substance can previously determine, for it is the whole current of the tension operating, or the relative tendencies of the molecules in all the substances exciting, that must decide which way the induced or excited tension shall have its pressure. Some may be of so positive a tendency as invariably to excite the positive tension, and some the reverse, and such will be known respectively as *positive electrics*, or *negative electrics.* Chemical dissolution may be made to keep on a perpetual and strong action more readily than mechanical friction, and may thus secure a continued stream of electric forces; and such stream must always have its two currents in opposite directions, for as the one polarity sets in one direction, the opposite polarity must counterwork in an opposite direction. The substance, as a wire of copper or soft iron, that takes these currents in their circuits is said to possess electrical poles according to the kind of electricity that forms the particular current, one the *positive pole*, the other the *negative pole.* The electric current is purely a stream of forces, and may not carry along in locomotion any of the molecules it particularly polarizes, any more than the wind carries along the standing grain-heads in the waves that pass over a wheat-field; for the molecules merely oscillate to and fro on their centres, in the alternations of the passing impulses of the two opposite currents in one and the same circuit, or in the alternations of induction and neutralization of one of the currents, each in its own wire.

But while the molecules of the substance thus alternately oppositely polarized, or alternately polarized and neutralized, do not move in their centres from their places, the electric force in its tension does move along the whole course of the wire, and like the wind over the grain, may carry along other things while it leaves in their places those things it makes only to oscillate. If the positive and negative poles at the ends of the two wires of a voltaic pile be tipped with each its piece of sharpened charcoal, and then these points of charcoal be brought in contact, they will immediately ignite. If then they be separated at a little distance, a stream of light will arch up from their tips between them, and by carefully guarding the eyes with colored glasses, in the intense light may be seen fine particles as dust carried across from the positive to the negative pole, and even slight detonations may be distinctly heard, as the particles of charcoal are torn off to fly through this bright arch. If the charcoal be displaced by soft iron balls, and the current be made to pass for some time, the ball at the positive pole will have its substance pretty rapidly carried away and be very considerably diminished in actual weight.

This is as the principle determines it must be. We are now able not only to say in theory that the molecules vibrate, and the torn off particles actually pass from the positive to the negative pole, but we can see this to be a necessary law, determined in the eternal principle that the central force must make its passage from positive to negative in the magnetic meridians, and when there is any derangement of polarity in any matter connected with the earth, this force must press itself upon that point of derangement and in its excess of tension must move through it, modifying in its course the molecular polarity and it may be, on occasion given, taking up particles of matter and carrying them along in its circuit across the chasm from one pole to the other.

It is to be remarked of this luminous arch, that it is not from any proper combustion of the particles carried through it, for the illumination is equally brilliant in the absence of all oxygen, and thus excluding all combustion. The light is from the decomposition of the material molecules which the electric force passes through in the atmosphere, and thereby liberates the heat that had been held in combination and which in its freedom becomes a new source of illumination. The electrical spark from a discharged conductor must be from the same determined law, and the zig-zag course of the lightning is of the same law of decomposition; the electrical force, in its passage through the material space-filling forces, being resisted and consequently deflected.

This principle of active circularity from positive to negative circumscribes many facts within its law, that have been clearly observed but not expounded. Among them is the different shape which the lights, or sparks, at the positive and negative poles assume; that at the positive pole taking on the diverging appearance as expanding from the polar point into a brush, while the negative pole gives off its light divergent, but as if radiating from the polar point as a star. This should so be from the order of movement. The positive electricity is the moving point of the tension, and in passing through any chasm in the conductor must in the resisting medium be expanded to a brush, while the negative point only as a static sustains the positive at rest in their meeting, and must have its light flattened and scattered to a star. And so also with the distances from the poles where the perforation through an intervening substance takes place. If a pasteboard be so placed longitudinally between the polar points of the wires that it shall meet the positive pole in one of its sides and the negative pole in the opposite side, the perforation of the pasteboard will ordinarily be directly opposite to the negative pole. The current makes its passage along the side of the pasteboard through the atmosphere the most readily, hence it passes on quite opposite to the negative pole when it must perforate the obstacle and join the negative force to balance itself. If this passage be made in *vacuo*, the current takes the pasteboard as its sole medium of communication, and makes the perforation any where between the poles in the most feasible place the peculiarity of the pasteboard happens to give, thus determining the law for the observed fact, that perforations in an exhausted receiver are seldom opposite the negative pole, but at some place between the poles and such point never beforehand determinable. The character of such perforations is also worthy of note. The orifice does not present the appearance of its having been forced through like the piercing of a bodkin, but as if it had been forced from the interior of the pasteboard each way out to the surface, leaving the edges of the orifice burred on each side of the board. If the movement of the positive tension had been a substantial movement, carrying along some moving body, which body had itself perforated the pasteboard, then must the orifice have been a pushing in on one side and a burring out on the other; but inasmuch as it is only the tension as force that moves, and this motion is by a perpetual oscillation of molecules, the molecules themselves become displaced in the pasteboard and the rupture must be as an explosion from the inside.

So also with the law of diffusion of the electric tension over the surfaces of conducting bodies. The induction is made while the conductor is insulated, and thus the insulating di-electric retains the electric tension to the surface of the conductor, and as this accumulates by the perpetuation of the exciting agency, there must be a continually augmenting tension on the conducting surface. This is found in fact so to be, and to be evenly distributed over a spherical surface, and to accumulate at the extremes just in proportion as the surface recedes from a spherical form. An ellipsoid has an excess at its extremities above that at its plane of the shorter axis just in proportion to its eccentricity, and a needle has almost the entire energy at its point, while all solids have the tension at the edges and angles in the same excess of their distance from a spherical centre. This must manifestly so be, since an accumulation upon any surface from whence there is no exit must range its positive tension on one side in opposition to, and in equilibrium with, its negative tension on the other, and thus the middle must be as an equatorial plane making its two hemispheres of electrical tensions, and which must proportionally spread themselves each over its own peculiarly shaped surface. In this is found all the practical utility of points in electrical conductors.

The electrical tension must always complete its movement in a circuit, and this circuit must ultimately have its two terminations somewhere in the earth. A conductor may be insulated, and the surrounding di-electrics may for awhile detain the accumulations to the conducting surface, but perpetual accumulations must ultimately force their passage to a further point of equilibrium, nor can the regressus stop in the perpetual accumulations, until the ultimate static be a balance for the originating dynamic. But this originating dynamic tension is always from the earth, in all human experience. The electrical machine and the voltaic pile have no permanent force, except as they are connected with the great reservoir of all electrical energy within the earth's surface. As the exciting agency goes on, and the molecular polarity is disturbed in the excited point, the tension from the terrestrial connection forces in, and the instantaneous opposite polarity arises at this point of disturbance and interruption of regular polarity. The terrestrial force may urge on to broader and broader surfaces, and all the play of the many-league surface of the thunder-cloud may be called into action, but nothing can balance the originating terrestrial force till the circuit puts its opposite end against the earth, and finds here the static exactly equilibrating the dynamic, and then all is at rest. The circuit may come completely round and find its return to the earth in the same connecting medium that was the passage for the tension out of it, as is the case in the common electrical machine, or it may run its points to and from the earth at many miles distance between, as in the telegraph-wire, but in no way can one end balance the other, nor any interruption of polarity restore its equilibrium, but by making the positive and negative stand somewhere in their extremes in the great magnetic courses of the earth. Thus, in fact, the electric tension completes the circuit, not by any imagined direct course beneath the surface of the earth from point to point, but only by falling at each end into the great circuit that goes up through the earth's pole and down the axis to the terrestrial centre. Each electrical circuit, whether through a dozen hands that hold themselves to the chains of the common electrical machine, or through the telegraph-wire that may span the Atlantic, is but a portion of the earth's magnetic meridian, taken up and turned and looped, and its polarities interrupted and used at pleasure; and such use of that very force which from the centre ensphereds the earth, may to any extent be made by man according to its necessitated laws. It is the same subjection of the powers of nature to human ends as that which employs the magnetic compass, and as the same force of magnetism when interrupted in its polarity becomes electrical tension, so that may be used like the magnet with reversed poles across the equator into an opposite hemisphere.

The combination of electricity and magnetism, whether as *magneto*-electric or *electro*-magnetic, has its clear determination of necessary laws from eternal principle, as truly as either one separately. The transverse polar action of the combined energies of magnetism and electric currents, which necessarily result from their principles, may readily give all the circular mechanical movements which experiments have disclosed. Thus the opposite helices of the electric current, that have been made to encircle the two bent arms of the soft iron bar, must make the modified molecules of the bar to take on the precise polarity of the magnet, and in the augmenting tensions of the electric current, we have the horse-shoe magnet of unrivalled energy. The principle of interrupted magnetic polarity will give all the principles of electric tension, and this determines all laws of electricity. The law as it comes out in nature is no arbitrary fact, that might as well have been otherwise; the fact is determined by a principle which is seen to be prior to it, and conditional for it; and only in the application of the principle can we have any scientific explication of the facts we find. We know not only so they are, but we know the reason why they are so.

6. Laws of Heat.—The principle of heat is found in the necessary radiation of vibratory movements, on all sides, from any point where the diremptive activity works adversely upon and in the midst of the antagonist activity. The same principle in general is the determination of both heat and light, and only as the heat vibrations are more divellent, and thus occurring in broader spaces and wider angles, are they to be found giving their phenomena dis-

tinct from the phenomena of light. In those particulars where the phenomena have the same laws, the explication will best be effected when we come to consider the laws of light, but in the instances of distinct law for heat we will here note the facts as determined in their principles.

A prismatic spectrum, which we shall more fully explain in connection with the facts of light, may be divided into portions of three different degrees in the refrangibility of the transmitted rays, the most refrangible being the most chemically efficient, the least refrangible the most thermally efficient, and the intervening the most luminously efficient. They overlie each other at their junctions, but the greatest thermal intensity is always low down, and sometimes beneath the luminous part of the spectrum. The slower, longer and broader vibrations have most momentum and are least refrangible, and these constitute the thermal radiations.

These thermal rays, when passed through an orifice by themselves, may be subjected to particular experiments, and the laws which govern them in the various particulars of reflection, refraction, diffraction, interference and polarization, are the same as for light, except as length and rapidity of vibration give the necessary modifications.

The radiation of heat is superficial only, inasmuch as the diremptive action beneath the surface of a body is retained by the antagonist working of the molecular forces with it; and such radiation is, as from the principle it should be, from every point in the surface of the heated body. This radiated heat is also taken and absorbed by other surfaces, so that the interworking is perpetually reciprocal, and no heat is lost. The body that gives less than it receives becomes proportionally hotter, and that which gives more than it receives becomes colder, but no body becomes so exhausted of its heat that it can radiate no more. The most intense pressure and violent friction may be given, and which will proportionally quicken radiation, but so long as the substance is held in combination it will have radiation still going on. The radiations are also reflected from polished surfaces and absorbed by rough and uneven surfaces, as in the case of light, and the same curvatures in each case give diverging or converging rays, and thus focal distances, under the same laws.

When rays of heat are absorbed by certain bodies or pass through other bodies, it is from the same laws as in the absorption and transmission of light, and yet as the light-vibrations are more refrangible than those of heat, it should be as it in fact is, that some bodies which readily absorb or transmit the one will be very slow in absorbing or transmitting the other. Bodies which readily transmit heat are termed *diathermanous*, and those which resist transmission are called *athermanoüs.* Bodies which are highly transparent may thus often be very feebly diathermanous. In no case is the same substance in equal degrees for both, the different refrangibility in each necessitating the distinction. Thus rock-salt is the most perfectly diathermanous of all bodies, transmitting about 92 per cent. of all the rays of heat which are incident upon it, while the rays of light pass with so much difficulty that the salt is simply translucent, not transparent. On the other hand, pure water is very transparent, but only very slightly diathermanous; and a plate combined of alum and green glass will readily transmit bright light while almost utterly impervious to heat. In general, that degree of refrangibility in heat which is also light will penetrate the same bodies, but the heat which is at the same time dark will be refused transmission by many transparent substances. The general principle for light and heat is the same, the modification in the law is from the modified vibrations.

When the light and heat are absorbed and not transmitted, they still remain in the body absorbing them, and this received diremptive action should make its correspondent effects in the body. That light which is less thermal in refrangibility will make its greater chemical effects, and that which is more thermal will manifest more heat, but it will be only as heat that the effects in dilatation will be exhibited. The retained heat is ever a diremptive activity, and except as combined in molecular antagonism so as to generate some new substance, it must loosen and separate the molecules from each other and thus expand the body proportionally to the thermal intensity. Thus as a fact solids expand with heat; and the infusion of heat perpetuated dissolves and isolates the molecules, and the body becomes fluid; a further absorption of heat and the molecules are still further dissipated, and the fluid becomes vapor. When the heat is so intense as to decompose the chemical combinations in the substance itself, and in this decomposition to set free also the diremptive force that had been held in affinity with the antagonist forces, and thus dissolving the peculiar molecules of the particular substance, we have *combustion*. The molecular structure of the substance is dissipated, and the liberated diremptive force is *flame*.

In this absorption of heat and dilatation of molecular

structure there is a peculiar law in all substances, that the passage from the solid to the fluid state by heat, and then again the passage from the fluid state to vapor, shall be accompanied by the using up and holding in a latent and imperceptible form a large amount of the heat imbibed, and which is known as *the latent heat* of *fusion*, or of *vaporization*. This also must all be given off again, in an open and sensible manifestation, before the vapor can return in condensation to the fluid, and the fluid to the solid. The chemical molecular structure is not changed, and thus there is no change of substance in the passage from the solid to the fluid and from this to the vapor, except as crystallization occurs, and thus no portion of this latent heat is used in any new chemical combination. The heat that had expanded the solid body only loosened and separated the molecular layers in the solid state, but for the fluid state it must loosen each molecule and quite surround and isolate it, and this it does by directly working against the antagonist or gravitating forces, and thus making itself to be balanced and neutralized thereby; and so much as is thus used is held, in the fluid, in a latent and imperceptible position. This must, of course, all be set free again before the fluid can become solidified.

And so in the fluid state, the molecules of the substance have been thus isolated by the combined action of the diremptive and gravitating forces, but the molecular structure has not itself been loosened; while now, in order that it may pass into the state of vapor, it is necessary that the forces which chemically form the molecules be relaxed and separated, though the chemical combination of the molecule is not dissolved but only loosened.

This necessary use of heat for the expansion of the fluid to vapor is, in the same way as before, a neutralizing and thus balancing of the diremptive and antagonist activities, which luxates and expands the combinations in the very chemical molecules themselves, and which in such neutralized action of course makes so much of the heat to be latent, and which must be set free before the vapor can be condensed again to a liquid. Different substances will use different degrees of heat for such fusing and vaporizing processes, but the law is determined by the very principle of diremptive action in the cases themselves. The latent heat of fusion and of vaporation is the heat statically used in separating the chemical molecules from each other, and also their elements among themselves.

And so with animal heat, the principle of the diremptive-working is clearly determining. The animal body as a living organism, is kept at a very steady temperature quite different from, and usually much above, the surrounding material heat. There must therefore be a continual evolution of heat going on within the animal organization. The temperature of the blood in the human family is about 89° Far.; of fowls, on an average, about 105°; of mammals, above 100°; and of fish about 70°, all of which is much above the usual temperature of the air or water in which they live. Experiments carefully made have proved, that the heat generated in animal life from respiration and assimilation, is analogous to that which is evolved in the combustion of a common candle. The watery vapor, carbonic acid, and azote given off had been supplied by the oxygen, carbon, and hydrogen, contained in the air and food, and the heat produced by the same quantity of materials corresponded in the animal assimilation and the lamp combustion. The life-force combines with the forces in the material elements used, and so decomposes the chemical molecules, that a portion of the heat is liberated in the lungs and sent on through the organism in the blood, while other elements in these chemical molecules are incorporated into the animal system. Our food and respiration perpetuate this supply for free animal heat, just as does the wick, tallow and air perpetually supply the heat from a burning candle.

7. Laws of Light and Luminiferous Bodies.—Although in the chapter on principles we found the place for that of light, after the formation of world-systems, yet inasmuch as the essence of light and heat are the same, and their principle differs only as the radiating vibrations differ, and also since none of the facts that may come under the intervening divisions of chemistry, crystallization, world-formations, and planetary motions, will have any special bearing upon the facts of light and their laws, it may be the most convenient and appropriate to give these facts and laws a place directly in connection with the facts and laws of heat.

The principle of radiating vibrations is the same as in heat already considered, and the sources from whence the vibrations radiate may be any point where the diremptive activities go out in excess of the antagonist forces. To the extent that the central energy shall propagate the vibrations outward and onward, will be the dimensions of the luminous sphere, and when the antagonist molecules equilibrate this diremptive force, the radiation will be exhausted. The great sources of light will be the central worlds of the systems, but any minor sources may be opened where any accumulations or liberations from decomposition of the peculiarly modified diremptions shall occur.

With no obstructions, the luminous source radiates equably on all sides and fills a certain sphere, and that sphere is limited from the same central source by the rarer or denser media that circumstances may compel the rays to pass through. If any obstruction occur, the rays are proportionally extinguished in it, and the shadow beyond is a perpetuation of extinct light in right lines from the source of illumination. If intervening material bodies permit the light to pass through with little extinction, they are known as *transparent* bodies, and if with so much extinction as, though still luminous, to destroy the capability of vision, they are called *translucent* bodies. Where the opaque body is of such relative proportion to the luminous source, that the rays from one side of the luminous body are extinguished by the opposite side of the opaque body, some considerable distance from the point where the rays of the other side of the luminous body become extinct, and the same thing occur in the whole outline of the opaque body, then the shadow will have its blended light and shade as a complete outside edge, and which is known as the *penumbra.* All of which fall under laws plainly determined in the general principle.

If the illuminating radiation strike the surface of a body at any angle of inclination, the rays not extinguished in the body will slide off or rebound, and still pass on in their turned course, notwithstanding the encounter, and such rays are said to be *reflected.* The principles of motion in the compounding of forces determine the laws of reflection necessarily to be as experience always finds them in fact, viz., that the lines of incidence and reflection are of the same angle, in the same plane, and on opposite sides of the perpendicular to the reflecting surface. Polished surfaces reflect with the least extinction, and curved surfaces, reflecting according to the law of perpendicularity to the surface, will give converging or diverging rays of all varieties, and open the whole field of optical laws in the different forms of mirrors.

The radiation that passes through a transparent substance will encounter differences of resistance in propagating the vibrations through the free ether and through the denser body, and in different substances the resistances will differ, but such resistance will turn the ray from its continuous right line, and this break in the line of radiation is called *refraction.* The compounding of forces in the principle of motion again determines the laws of refraction to be as the facts are always found, viz., the angles of incidence and refraction are in the same plane; and the angle of refraction is the same in all cases for the same substance.

A polished curved surface modifies the direction of the refracted ray according to the curvature, just as above in reflection, and this opens the field of optical laws for all differently formed lenses. The different refrangibility in the parts of the same lens will bring some rays to a different focal point from others, and this scattering of the rays away from one common focus is termed *aberration.* The correction for aberration by applying another substance to the lens of a proportionally compensating degree of refrangibility is guided by the same necessary law as that which necessitates the inconvenience.

A succession of molecular vibrations in a right line out from the luminous source is a *ray*, and as the eye must take in many such lines of molecular vibration at each moment of vision, the combination of many such single lines is still called a ray, and of course admits of an analysis to just the extent that single lines have been put together in synthesis. We have thus single molecular *lines* of light, and combined lines of indefinite numbers in *rays* of light, and many such rays combined are a *beam* of light. A prismatic medium receives on the incident surface a variety of rays in all their differences of vibration, both in rapidity, and in breadth of prolate and oblate expansion. Their different refrangibility must be determined by their difference of vibration, and the reaction of the prismatic medium upon them in both its angular expansion of surface and inherent substance. These rays will thus come out from the emergent surface separated and distinguished according to this difference of refrangibility. The slowest and longest vibrations will have been the least refracted, and the quickest and shortest the most refracted, and if these prismatic rays be now received upon an even surface, they will range themselves according to this discriminated variety of vibration, the slowest and longest the nearest to the edge that gives the refracting angle, and the quickest and shortest away from it, and the proportionally refrangible rays filling their proper places in the space between. This illuminated space is known as the *prismatic spectrum*, and the principle of diremptive action determines, in the conditions, the law of the spectrum, viz., an oblong illumination with its discriminated vibrations, and in this its colors. The lower extreme is red, and the upper is violet, and from the red up to the violet are the orange, yellow, green, blue and indigo. The analysis can be carried no further, for another prismatic refraction makes no more separations. Two blended colors, as the yellow and green into a blue, may stand over against a primary prismatic blue of the same intensity, and the prism will decompose the first but leave the last unchanged. By the interference of vibrations their lengths and rapidity may be calculated. The transverse vibrations are for violet about 170 ten-millionth parts of an inch, and for red about 260 ditto. The number of vibrations progressively in an inch are for violet about 58,000, and for red about 38,000. The number of vibrations in a second are, for violet about 700 billions, and for red about 450 billions. This rapidity of vibration, though inconceivable in detail, is quite comprehensible in the general principle. The diremptive action is from the centre of the luminous sphere out every way in the equatorial plane, and thus across each hemispherical layer. In crossing each layer its oscillation must give vibration to each molecule it possesses, and the equatorial plane being always a plenum, the central pulsation must be felt at once to the circumference and thus through all the layers that the equatorial plane crosses, and which would be instantaneous but for the elastic compressibility. The rapidity of propagated molecular vibrations is therefore only so much less than instantaneous, as the plenum of luminous ether is successively compressible.

The prismatic spectrum, thus wholly according to the determination of the great principle of the light-force, gives also a *chromatic aberration* from necessary laws, when the rays of the spectrum pass through a lens. Not only is the lens differently refrangible from peculiarity of curvature and substance, but each color of the spectrum is also different in refrangibility, and must make its dispersions in the focal divergency, and also on the colored plane. One color will take a different position from another, and all images must thus be confused within and fringed without, except as corrected.

When two luminous sources have their spheres in such a way as to cut each other, the lines of molecular vibration must cross and thus interact upon each other. Where these vibrations act from each other they must intensify, and where they act upon each other they must neutralize the vibrations. Such interaction of vibrations gives luminous *interference.* The result must be a series of alternately intensified and obscured light, and in the extremes, bright light alternating with total darkness. Many peculiarly interesting optical phenomena result from this law of interference, and are expounded by the principle which determines it.

So also, if rays of light pass the edge of an opaque body, the shadow instead of being exactly defined in a right line, will have a penumbra in a diverging pencil from the edge of the body. The vibrating molecules in their radiation past the edge are reacted upon by the obstruction, and a new circular vibration commences in it, and forms itself about it, and goes on enlarging from it, as would outgoing waves of water when passing the edge of an obstruction. This diverging penumbra is called an *inflection* or *diffraction* of light, and has all its laws determined in the great principle. When the ray of light passes through certain peculiar crystalline bodies, in a certain direction, it becomes entirely divided, one part passing through without refraction, and is called hence the *ordinary* ray, the other part is more or less refracted and called the *extraordinary* ray. As the prism divided the ray into the different colors of the spectrum, this crystal substance divides it into two rays, homogeneous in color but distinct in planes of polarity, and which is known as *double refraction*, and which will be more fully explicable from what follows.

The two alternate vibrations prolate and oblate of every molecule, by the diremptive action, give occasion for vision and color from each of the double movements. If then, any arrangement turn the molecular vibration on one side, and the ray is made to meet the eye with one vibration veiled and lost in its transverse movement, there will be but the two sides of the vibration luminous, and the two ends will be in darkness. The ray being thus luminous only in two opposite sides is said to be *polarized*. This may be effected by so turning the vibration in a series of refractions and reflections in two mirrors placed to each other at certain angles; or by making the ray pass through certain crystalline substances which destroys one kind of vibration or combines both in one; and also by the peculiar double-refraction above noticed. In the last case, the ordinary ray has one plane of polarization, and the extraordinary ray another, when the consequent interferences of vibration between them darken to each other their opposite neutralized action. If then, in any of the above cases, the ray be made to turn itself round, by means of a revolving mirror, it will be alternately light and dark on opposite sides, and partially luminous in the places between. If this double refraction be made to occur in exceedingly thin plates of crystal, the ordinary and extraordinary rays will be somewhat blended by their lying one upon the other, and their passing through the refracting medium at different rates of velocity will bring them to an interference of vibrations, and this, with their opposite planes of polarity, will give all the beautiful and wonderful phenomena of chromatic polarized light, exhibited in some experiments with varied axial and acute prismatic crystals.

But beside these facts and laws of light itself, there are some that may be considered as determined in the principle that makes the central bodies of the systems to be luminiferous.

The facts connected with the changing appearances on the face of the sun are specially of this description. A careful observation of the sun through a telescope will ordinarily discover more or fewer dark or colored spots upon the sun's disk, and these spots at times become very numerous and very large. In one case, by mensuration and calculation, an observed spot was found to cover a space on the sun's disk 46,600 miles long and 27,960 miles broad. They have a determinate region on the face of the sun where they appear, and are not found within an equatorial belt that may be imagined like the torrid zone to encompass the sphere, but always on each side of such belt or zone for a considerable distance, and then beyond these two separate belts towards the polar regions the spots again do not appear. They are of all shapes and sizes, and at times have a very dark central spot with a surrounding region of blended light and shade, and at other times the whole spot is partially luminous, and though considerably darker than the main face of the sun, is still of a faded twilight appearance. The darker spots evince in various ways that there is a deep opening in a luminous envelope about the body of the sun, and that this opening is broad at the surface and shelving down its sides by an inclined slope to the dark bottom on the body of the sun itself, like an immense crater of a volcano. These funnel-formed openings often rapidly change their shapes, and sometimes streaks of light stream across them from side to side, and such streaks rapidly augment and fill up again the whole opening. Around the central darkness there will often be found branching arms that run out from its margin to long distances within the surrounding illuminated portions. Beside these positively dark spots, the contiguous region often appears of a waving or mottled aspect, having very irregularly defined outlines, and rather as a veil of thicker and thinner light thrown over broad portions of the sun's face. The luminous matter of the sun's envelope is thus evidently seen to be a moving and changeable covering, sometimes rarified and sometimes completely broken through in pretty well defined regions, and other portions of the sun constantly maintaining their brightness, and over which the luminous envelope is never ruffled or broken. Whatever may be the source of agitation and disruption, the effects only appear about equi-distant from, and parallel to, the equator of the sun.

These facts may very well be occasioned by the manner in which we have found the diremptive force must work in throwing a luminous atmosphere about the central suns of systems. The pressure of gravity in the surrounding ethereal matter, in which the diremptive activity is so balanced that the heat-force is latent, disturbs and destroys the balance and forces the diremptive activity into exercise, and thus the latent heat becomes manifested heat. The diremptive activity is against the solid surface of the sun, and out from it into the balanced ether, in a perpetual disparting or divellency of agency that renews or restores the vibrations which had ceased from the equilibration of forces. This renewed radiation of the vibrating movement, intensified by all the force of gravity that is induced by the quantity of matter in the sun, becomes so much quicker and sharper that it rises to a luminous state, and is heat elevated to light. The action upon the sun and out through all the surrounding ether must be according to the only principle of diremptive activity, by a polar and then an equatorial movement alternately, and thus a continual oscillation from a prolate to an oblate form through all the successive outlying spherical layers, so far as the radiations shall penetrate. This action upon the superficial body of the sun in its rotation must accumulate the luminous atmosphere about the equatorial region, and send it off each way from thence toward the poles.

When then we conceive of this luminous atmosphere as itself an imponderable fluid, and working off into the surrounding ether by its radiating vibrations, and perpetually supplied from the latent portions in the ethereal mass which its gravity is constantly pressing up to the sun's surface, it is readily seen that both the diremptive activity and the sun's rotation conspire to induce just such phenomena as these solar spots, and in just such general localities on the sun's surface. The equatorial portion or torrid zone of the sun will be the fullest and steadiest supplied, and the currents sent off from this zone, both by the diremptive force and the action of the sun's revolution, must necessarily flow in eddying streams out into the two temperate zones on opposite sides, and might be expected at times to occasion there just such partial openings and deep rents and gaps in this luminous cover of the sun, as we find actually to disclose to our observation more or less of the solid and unradiating body beneath it. Such openings in the illuminated covering may last for a number of the sun's revolutions, and thus they will appear to move over the sun's disk, disappearing from one limb and again appearing at the opposite. These eddying currents must spend themselves in these zones, leaving the polar regions perpetually again undisturbed and luminous. What we might almost predict, we may pretty safely explain, from the necessary circumstances.

The great luminiferous bodies are thus the central suns of the separate systems, and which become self-luminous stars to each other, while the planetary bodies and their satellites have only the sun-light radiated upon them and reflected from them. When the radiations or reflections are cut off, the planets are forthwith opaque.

But there are many other though far smaller sources of light, independent of all irradiation and reflection, than the sun of our system and the distant stars. All direct light from self-luminous sources may be conditionally reflected or refracted, and the rays revolved so as to hide by turns one plane of the vibrations, and thus to become polarized; but reflected planetary rays so lose the regularity of their planes of vibration that they cannot become polarized. The fluid ether and the solid substances of matter have much diremptive force or heat combined, or latent within them, and whatever liberates this heat and sets it again in free action, opens anew the sources of radiating heat and light, and makes the dormant to become apparent. This is done in many ways. The violent dissipation of the substance in combustion; the rapid destruction of the substance in chemical decomposition; the dissolutions of the electric and galvanic currents; all these liberate the confined and balanced heat-and-light forces, and the light streams forth again as from a new fountain. All combustible matter is such on account of its component diremptive forces, and the combustion that sets them free induces at once the phenomenon of flame, and that matter in combustion is thus luminiferous with the primitive light that may be polarized. Chemical and electrical light are alike decompositions and thus liberations of the primitive diremptive forces. No portion of the diremptive force, that has ever proceeded from the great universal centre, has gone up again through that centre, and been received as pure spiritual activity by the Absolute, nor can it otherways be lost in any outward dissolutions and annihilations. All light and heat that ever was, yet is, and wherever bound in composition with other forces, may again become decomposed and free, and hence new streams of light may be made anywhere to flash out around us. All matter, it may be, is in this sense luminiferous, that it holds in composition something of the light-force, and which is therefore decomposable, but it is only while the passing decomposition goes on, and the freed force makes itself to appear, that we call the body light-bearing or self-luminous. The burning lamp, the lightning flash, or the slow chemical decomposition of putrid fish or rotten wood; all are so many instances of the light-forces which had ceased to shine because locked in composition with other forces, but which now stream forth anew because once more set free to push their circling vibrations through every point about them.

The fact that flame is fed by combustibles, and extinguished by non-combustibles, has its law in this, that only as the decomposition goes on can the light-force be liberated. Water is the great extinguisher of flame while it is itself undecomposed in the flame, but let the diremptive action be sufficiently intense to decompose the water, and it will add the diremptive forces in its own substance to the flames, and make them to glow with greatly augmented brightness. The extinguisher of flame becomes thus fuel to the flame, so soon as the dormant light-force in itself is set free in the decomposition of its own substance. In the same way, spontaneous combustion breaks out in new flame, when some chemical decomposition has let loose the diremptive forces that had been held inactive by superior counter-agencies.

8. The Law of Chemical Forces.—Matter is given to us in masses; chemistry analyzes the masses, and gives to us the various chemical substances. These varied substances when further analyzed, are reduced to some three-score (61) so called simple substances, because the chemical solvents have not sufficed to carry the analysis any higher. These simple substances, as now considered, may be indefinitely diminished by further more successful decompositions. But how many soever the simple sub-

stances may be as resisting further chemical analysis, the great principle of all chemical synthesis in its original combinations, is that the antagonist and diremptive activities are the only simple forces, and that by the varied counter-working of these upon themselves, in direction, numbers, and energy of the simple molecular forces, the first chemically indecomposable substances are formed. These primitive substances, more or less, are then chemically combined to constitute all the distinctive compound substances in nature, whether earths, metals, or gases. Not at all that dead inert atoms are brought into juxtaposition by some assumed forces, the atoms being the things considered and the forces only hypothetically imagined, but the intelligently apprehended forces both make and move the atoms, and thus it is only for forces that a true philosophy has any interest.

This general principle of the combination of the two simple forces in various proportions and directions, to constitute what has been improperly understood as truly simple chemical substance, has determined a remarkable and universal law of the action of chemical affinity. To mark this determining principle the more distinctly, let it be noted that the parting from any given substance in composition, and coming together with some other substance in combination, is known as *chemical affinity*, and the degree of energy with which the separation and new combination is effected, is known as *the force* of chemical affinity. All cases of complete combination of the liberated substances give what is known as *the definite action* of chemical affinity, and those cases where one substance is only dissolved in another, and there continues a solution but no completed combination, are known as exhibiting only *an indefinite action* of chemical affinity. The law above referred to is, *that in all cases of the definite action of chemical affinity, heat is evolved.*

The converse law—that in all cases of indefinite action of chemical affinity, cold is induced—has nothing that need here to be remarked, as its determination is seen in what has before been attained in the latent heat of fusion, and of evaporation. Thus when the solid becomes a fluid, or the fluid becomes vapor, from the force of heat, much of the applied heat-force must be used in separating and isolating the distinct molecules, and as just balancing and equilibrating itself with the molecular forces it beomes fixed and is insensible or latent heat; but as taken from other surrounding substances and thus fixed, it has induced sensible cold, or the absence of so much sensible heat.

But, that all definite action of chemical affinity evolves heat is of much significance, as being determined directly from the principle of the combination of the simple antagonist and diremptive forces to constitute the so called simple chemical substances. As an example, we may take 1 lb. of hydrogen of specific heat 3.2936, and 8 lbs. of oxygen of specific heat per lb. 0.2361, and the compound will be a watery vapor of 9 lbs., with specific heat per lb. 0.8470. The mean would be—

$$\frac{3.2936 + 8 \times 0.2361}{9} = 0.5758.$$

The difference then will be 0.8470—0.5758 = 0,2712. This amount of heat evolved in the combination must have come from its liberation out of the simples hydrogen and oxygen, thereby showing them not to be truly simples, but compounds of the simple forces.

In another instance, we may take 2 volumes of nitrogen and 1 volume of oxygen condensed to nitrous oxide of 2 volumes, and the resolution of the nitrous oxide should use up heat thus made latent and thereby induce cold. But instead of cold, if wood-charcoal be burned in the nitrous oxide, the combustion gives 19623 units of heat, and burned in pure oxygen the combustion gives 14544 units of heat. The nitrous oxide has gained this excess of heat over the oxygen, viz. 5079 units, beyond the sensible heat that must have been lost or made latent. This can only have come from its liberation in what is termed the chemically simple nitrogen.

So in all combustion, the dissolution of the substances disengages or liberates a great amount of heat, without any condensation and usually with a large expansion of volume, as of nitre with charcoal. This law of evolving heat cannot be determined from the setting free of any latent heat of fusion, but is determined only in the principle that the really simple forces are combined to form the so called chemically simple substances, and the decompositions of the chemical atoms themselves in combustion give off their constituent portion of the heat-force.

These simple substances combine and form their various compounds, and in the combination the simple substances so change their action that the compound is a third thing wholly different from either of the separate ingredients. In their analysis the simple substances again appear and the new substance as a compound is lost. Thus oxygen and hydrogen, brought into combination, form the distinct thing, water; and hydrogen and nitrogen combined give ammonia. An analysis of sea-salt gives the simples chlorine and sodium, and potash analyzed gives oxygen and potassium. The compounds are as different from their elements as these elements are different from each other. Nothing, in fact, is truly a definite combination that does not result in a new third substance. The affinity of salt and water acts only indefinitely in securing a solution but no combination. The salt and the water still remain in their unchanged substance, and no third thing is given. The atmosphere, which contains 21 parts oxygen and 79 parts hydrogen is still rather a conjunction than a combination of the elements, and the action of the affinity is indefinite, since atmospheric air is not a new definite substance. The primal forces which, in their modified working, constitute the chemically simple substances, change their working, and as above seen some of the diremptive force or heat is liberated, in their combination into a new substance.

And still further, the principle of antagonist and diremptive activities is seen in determining the laws of chemical combination, in that not only must there be definite affinities, but these affinities must always be in specific proportions of the simple substances. The forces cannot blend and stand to each other as static equivalents except in certain specific degrees of relative energy, and such simples in such proportions are known as *chemical equivalents.* The quantity in energy of molecular force, or as the same thing, of weight, is thus always, in chemical combinations where the simple substances may produce more than one distinct compound, in one of the two following ratios of the first substance to the second, viz., 1 : 1 1 : 2 1 : 3 1 : 4, &c.; or, 1 : 3 1 : 5 1 : 7 &c. Some of these chemical equivalents are as follows: Hydrogen 1; Carbon 6; Oxygen 8; Nitrogen 14; Chlorine 35,50; Mercury 203.

Thus, hydrogen and oxygen combine in two proportions, and in each form their definite substance. 1 hy.: 8 ox.=water. 1 hy. : 16 ox.=deutoxide of hydrogen. Nitrogen and oxygen unite in five compounds, and all in their relative proportions only, viz. 14 nit. : 8 ox.=protoxide of nit., or laughing gas; 14 nit. : 16 ox.=deutoxide nit.; 14 nit. : 24 ox.=hyponitrous acid; 14 nit. : 32 ox.=nitrous acid; and 14 nit. : 40 ox.=nitric acid, or aquafortis. The same weight, or force, in nitrogen is to the weight or force of oxygen in these definite affinities respectively a chemical equivalent as 14 : 8 : 16 : 24 : 32 : 40. Carbon and oxygen again have two compounds, viz., 6 : 8=carbonic oxide; and 6 : 16=carbonic acid. Mercury and chlorine have also two compounds, viz. 203 : 35,50=calomile; and 203 : 71=corrosive sublimate.

The necessity for this universal law in chemical equivalents is seen in the great principle of the working of heat in the primitive ether. Where it only separates the ethereal molecules and perpetually forces itself in among them, it makes the whole a fluid mass ultimately so comparatively dense that the new diremptive current coming in drives the old before it, and the chaotic matter is wheeled into worlds and systems. These primitive forces then combine, in what are chemically simple substances, by such proportions of energy as are indicated in the chemical equivalents, they respectively possess; and such proportions of energy in combination constitute the *chemical molecule*, or atom, and these so-called simple chemical atoms combine, according to their affinities and equivalents, and form the many chemically compound substances. The separate working of both antagonist and diremptive forces in the simple chemical atoms determines that the atomic combinations must be in their definite ratios, and that all such atomic combinations must liberate heat from the modification of the atoms themselves.

The aggregation of the chemical atoms in masses gives material bodies of such homogeneous or mixed matter as the nature of the atoms determines, and while the force of heat has been put in combination with the primitive ether in various ratios to constitute the chemical atom, and which heat in combination is insensible, there is also in all such material bodies more or less heat between the chemical atoms, which in its diremptive working keeps them apart, and this is sensible heat. The more heat diffused the more the body is expanded, and the more the heat is abstracted the smaller the volume of matter becomes. If the substance be in a fluid state, the cooling process contracts the volume proportionally till the abstraction reaches and begins to exhaust the latent heat of fusion which isolates the atoms and keeps the mass fluid, and in the exhaustion of this the chemical atoms compact themselves by a centripetal pressure, and usually the fluid suddenly contracts greatly in volume and the whole is *solidified.* The amount of latent heat of fusion differs widely in different substances, according to the proportions necessary to isolate and make fluid the chemical atoms. Thus zinc is 49,43 units while lead is but 9,27 units, and therefore the shorter process of solidifying in melted lead than in zinc.

The amount of latent heat in water is as much as is needed to raise its own quantity by weight through 142° 65 Far., and which in its excess makes freezing water so gradual in the process of solidifying. In the same matter, the time for liquefying and solidifying is alike, for the same amount of latent heat is given and taken away.

9. THE LAWS OF CRYSTALLINE FORCES.—Fluid matter on cooling to a less or greater degree in most cases suddenly becomes solid, and the solid body takes on the form occasioned by gravity or pressure. But in the case of some matter, fluid or held in solution, there is a sudden solidification which takes on its peculiar form wholly irrespective of gravitation or any external compression. If salt or alum be dissolved in water, the evaporation of the water, or the immersion of some solid substance, brings the dissolved ingredients to a solid state, having peculiar mathematical forms of regular edges and surfaces and solid angles, and we term the process of such solidifying *crystallization*. The crystals for the same salt, or other solution, have always the same form in edges, sides and solid angles. Even if an already formed crystal be broken in fragments of divers shapes, and these be so introduced as to become the nuclei for further adhering crystallizations upon their own substance, all the fragments will at once accumulate upon themselves an augmented crystallization of the same old form as the unbroken crystal had presented.

These forms which different substances take on in crystallization are some dozen geometrical solids, and which may be classified into six different kinds, constituting a complete system of crystallography. If one take up any philosophical treatise on crystallography, as Draper, or more elaborate and complete, as Dana, he will have all the forms given in regular diagrams, and ranged under the following systems determined by the peculiarity of their axes. The axis is a line imagined to be drawn through the centre of the crystal, and around which all its parts are symmetrically disposed. There will be a number of such axes in all crystals, and according to the difference of geometrical form will be the difference in the number, comparative length and direction of the axes. The modification of the axes is thus the basis of classification in crystallography.

There is, 1st. The Regular, or Monometric system; having three equal axes at right angles to each other; and containing the three geometrical solids of the cube, the regular octahedron and the dodecahedron. 2d. The Square Prismatic or Dimetric system; having three axes intersecting each other at right angles, and the vertical of unequal length to the two equal lateral axes; and containing the two solids of the right square prism and the right square octahedron, each having two varieties of axes, one terminating in the centre of the sides and the other in the middle of the edges. 3d. The Right Prismatic or Trimetric system, having three axes of unequal lengths and right angled intersections, containing the right rectangular prism, right rhombic prism, and the right rectangular based and the right rhombic based octahedron. 4th. The Oblique Prismatic or Monoclinic system, having three axes all unequal, two intersecting at right-angles and the third oblique to one and at right-angles to the other, and containing the oblique rectangular and oblique rhombic prisms, and the oblique rectangular based and the oblique rhombic based octahedrons. 5th. The Doubly Oblique Prismatic or

Triclinic system, having three axes all unequal and all with oblique intersections, containing the two varieties of the doubly oblique prisms, and of the doubly oblique octahedrons. 6th. The Rhombohedral or Hexagonal system, having four axes, three of which are equal in the same plane and intersect at an angle of 60°, and the fourth or principal axis perpendicular to all the others, containing the hexagonal prism and the rhombohedron.

These may sometimes be imperfect on one side or even wholly wanting, and are thus *hemihedral;* and sometimes the same substance may be in double or twin crystals and be *paragenic*, if originally so formed, or *metagenic*, if so formed after the commencement of crystallization; and also the same substance, from difference of temperature or other cause, may crystallize in two kinds, and run into two systems, and be thus *dimorphous.* There may also be sometimes the meeting of opposite faces and thus the turning of the axes in opposite directions, which will give to the crystal a knee-shaped or *geniculated* form; but ordinarily the same substance crystallizes in the same form.

Now, if we consider the combinations of the antagonist and diremptive forces in the chemical molecules when passing from the fused or liquid state to the solid, to change their activities according to their modified substances into the directions of these intersecting axes, we shall have so many varied forces, and which must each one build up its own crystalline form according to the axial direction and degree of energy. The cube will necessarily be engendered by a perpetual working, each way from the salient nucleus, of a force equally antagonizing in the direction of the right-angled intersection of the three axes in the first system; or if it regularly diminish in action in the direction of the two horizontal axes and continue constant in the perpendicular axis, it must build up the regular octahedron, of the first system; or if it regularly diminish its action in the direction of all the axes, it will continually cut off the twelve edges of the cube and necessarily ultimately complete the dodecahedron of the first system. And so, by a change in the action in the working force, according to the direction of the axes, and in regularly modified degrees, every form in all the six systems will necessarily be engendered from the central point or nucleus. The forces will determine the forms of the crystal, and at each point will constitute the space-filling molecule, and thus be both molecule and generating crystalline force in one. The supposition of dead atoms of different shape and specific attractions is both superfluous and contradictory, for the passive atom is helpless without the generating force and wholly useless with the generating force; such force truly doing all the work, and constituting all the substance of the completed crystal.

And still further, this method of generating the crystals must determine also the laws of *cleavage*. Crystals are more or less easily split into their laminæ that lie one above another, and between which there has been in the generation of the crystal a separating or discontinuing of the substance. The general integrity of the mass is thus in certain planes interrupted, and one leaf lies disparted from another. The facts are that in the same species of crystal, the cleavage gives the same forms and angles. The lines of cleavage are parallel to each other. Similar forces cleave from each other with equal readiness. The cleavage of similar planes has the like polish or lustre.

Such facts are all in full accordance with the idea of crystallogeny, for as the forces successively range their atoms in faces that must be determined by the order of working in and from the axes, so the successive accumulations of the matter arranged by the forces must be superimposed upon itself, layer after layer, in the same direction, and the equal action in generating each layer will make both their separations and surfaces and polish, or otherwise, to be similar each to each. Crystals must, therefore, have a cleavage in several different directions, in some of which the separation will be more readily made than in others, but always in the same directions will be of the same general characteristics. The determining force of generation must determine also the cleavage.

But the determining principle carries us very much further in reading the necessity for the laws in crystallography, and even forces to a consilience within the law some phenomena hitherto held to be quite anomalous.

In ordinary solidification usually the matter suddenly and largely contracts on the abstraction of the latent heat of fusion. Mercury especially suddenly solidifies at— 38° 8 Far., the column in the tube of the thermometer sinking into a solid far down within the bulb. But, on the other hand, all crystalline solidifications expand, and some very greatly at the moment of crystallization. Water particularly expands in congelation to about one-seventh of its own bulk. The method and progress of expansion is very peculiarly marked. From any point of temperature above, on the abstraction of heat, it contracts regularly till it reaches 38° 8, remaining about stationary for 1° or 2°, and then on further loss of heat regularly expanding till it comes to 32°, when commonly congelation occurs, and the expansion to one-seventh of the volume immediately succeeds. Very careful cooling without agitation may delay congelation down to 22° or even lower, and the regular expansion goes on; but on congelation from a lower point than 32°, the temperature at once rises to 32°, and the expansion in congelation is so much less at the lower degree as had been gained in the passage from 32° to it, so that the whole expansion in congelation is the same one-seventh in all cases.

Now, why do crystals expand, when all ordinary solidification contracts the volume of matter? The natural conclusion might be, that on this abstraction of the large amount of latent heat there would follow a large degree of diminished volume. The beneficial results of expansion in crystallization are of very great moment. The escaping of such an amount of latent heat gradually, and giving it free and sensibly in the surrounding fluid, makes freezing to be a gradual process, giving warning and time for any needed precautions. And then this large expansion forces the congelation to the surface and throws the evolved latent heat into the water below, and thus preserves the streams and lakes from entire solidity. But has the benevolent end been reached arbitrarily, without or against an immutable principle? The grand idea of diremptive activity, as we have attained it, enables us to see that the principle determining this law of the crystal, is as eternal as that which has conditioned any other law of nature. The benevolence is seen not in violating principle and making arbitrary laws, but in applying principles intelligently, and putting the right law in the right place; filling the streams and lakes with a fluid that crystallizes, and not with melted oils or metals that condense in solidifying.

The large quantity of latent heat in water secures the slow process of congelation by its gradual liberation, and as that only sufficed to just isolate the chemical atoms, and give occasion for a mutual flowing of each over the others, thus making the mass a liquid, without decomposing the chemical atoms of either the oxygen or the hydrogen, or dissipating them in a vapor, so the evolving of this heat has not given any occasion for contraction as its latent presence gave no occasion for expansion. It held itself just balanced in the midst of and around the atoms, but did not crowd nor dilate them. These chemical atoms, however, of 1 part hydrogen and 8 parts oxygen, have their own natures from the forces which compose them, and which must conform to the laws that the great principle of their constituent antagonist and diremptive forces has put within them, and so soon as the latent heat that liquefies them is withdrawn, they must act upon each other in such compensating affinities as shall turn their direction, and form their axes, and push out their atoms to the cleaving surfaces of their peculiar geometrical solid. If their polar direction had been simply in one axis, as in the mere antagonist-working, they would have hardened directly in towards the centre like the solidifying of uncrystallized bodies, but now they have their many poles and axes, and they must harden in the direction of their corresponding plane surfaces and solid angles and edges. A consolidation towards a central point must contract the volume of the mass, from the nature of the case, since it is an ensphering process, and the bringing of every chemical molecule to stand together with its fellows in the least possible space. Such solidifying must secure contraction of volume in the escaping of the latent heat-suffusion.

But the crystal hardens in geometrical solids, and the polyhedron must have more space for the collocation of the same atoms, and the varied surfaces with their lines of cleavage must lessen the density and enlarge the bulk, and thus the crystal cannot take on its solid form without more or less dilatation, and the amount of expansion will be determined from the form and structure of the solid and the force of the polar activities. When the cooling has reached 38° 8, the latent heat begins so to pass from its neutralization, that the intrinsic polar activity may also begin its agency, and though the fluid state maintains its mobility till 32° and lower, yet will the polar adjustment together with the waking but not yet escaping heat, begin their dilatation, and only finish the one whole work when they have carried the fluid to its complete congelation.

The same principle, partially carried out in a sub-crystallization, must determine the solidification in cooling of cast-iron, bismuth, antimony, and some of the alloys, as bell-metal, which may take shape from a mould in casting, since there is a small expansion in cooling, while most metals contract and shrink from the mould in coming to their solid state. Where there is a multipolar pressure in solidifying there must be enlargement, and where there is only uniaxial action and thus spherical concentration there must be diminution of volume. There may be the widest differences in the points of temperature where solidification takes place; this will determine nothing in dilatation and contraction. Mercury congeals at 38° 8; spirits of alcohol at 152°; and English wrought-iron is still solid at 2912°, and platinum at 3082°, and though carbon must have been fused to harden in the diamond, yet does this crystal maintain its solid state against the highest heat that can be artificially applied. Not the point of melting, but the forces that interact in cooling, determine both the fact and the form of the crystallization, and the fact and the degree of dilatation. The principle of diremptive action necessitates the laws of all crystallology.

10. THE LAWS OF WORLD-SYSTEMS IN THEIR ARRANGEMENT AND MOVEMENT.—As the stars, which may be assumed to be the suns of other systems, are seen from our terrestrial stand-point, they appear of varying magnitudes, and the numbers increase as the magnitudes diminish. If it be taken as a general fact that the smaller are the more distant, it will be readily apprehended how two stars of unequal magnitude may often appear as joined one to the other, and thus presenting the phenomena of double-stars. But as lying in nearly the same line of vision, and only appearing in contact, or nearly so, while one is at a vast remove beyond the other, such cases give double-stars only as *apparent*, and the two bodies have really no special connection one with the other. More than 6,000 such apparent double-stars have been observed, including both hemispheres, but they have nothing remarkable in themselves more than any stars of the same unequal distances, except as these happen to lie in nearly the same lines from our point of observation.

But there are double-stars that are not only found apparently joined in two contiguous beams of light, they also gradually and regularly vary their mutual positions, and thus manifest that they are physically connected in some common centre of gravity, and may be known as *physically* double-stars. More than 650 such physically double-stars have been observed, having a relative motion with each other, and not merely in relation to other stars by parallax of our or their change of position. Of these, 16 have had the elements of their orbits determined, and some have completed more than one revolution since their discovery, and have even presented the striking phenomenon of one so-called fixed star in occultation by another. Their periods of revolution are found to differ very largely, from 30 years to 630 years. Their immense distances, and especially their shining by their own and not by reflected light, inasmuch as their light may be polarized, determine them to be true stars and not any cases of planetary revolution.

Now, the laws of such double-stars, revolving the less about the greater, or both about a common centre, are readily determinable from the principle of their construction as given in Chapter II. Sec. 16. They were produced by the parting of one stream of the diremptive force, turning one part about the other, and the motion thus given and its velocity must continue constant in their subsequent condensing and throwing off their planets if they have them. Their difference of gravity, and propagation of their radiating vibrations through each other's ethereal spheres, account for the changes of color so remarkably observable in them.

But inasmuch as the worlds which compose the solar system are comparatively near to each other, and within the reach mostly of distinct telescopic observation, it may well be supposed that the opportunity must here be afforded for furnishing the greatest number of instances where the determining principle and the thence determined law shall come into full accordance. Our solar system can alone come within our observation, for its planets and their satellites are the only ones that can be reached by the power of any glasses yet invented. The fixed stars are as so many single and solitary worlds, with the comparatively few cases of known double-stars as above, which have their own relative motions, and it is only from analogy that we conclude that any of the millions of fixed stars are also suns with their revolving planetary worlds about them. The principle of the formation of world-systems induces the expectation, that the diremptive force will be sufficient, in the vast majority of cases, to occasion an excess of tangential or revolving force above the adhesive or gravitating force, and in consequence to drive off successive planetary masses; but the facts, if such there be, come within our experience only in our solar system. We shall here, however, find many conformities of determining principle and determined law, and which must have equal validity in all analogous cases.

When the volume and the mass of any world is determined, dividing the mass by the volume will give the *density* of the matter contained; and in the bodies of our system there is a general though not an exact and uniform increase of density in the matter of the worlds, from the exterior to the interior bodies. Taking the density of water as the common standard of measurement, we have the specific densities of the planets successively as follows:—

Neptune,	0,97	Mars,	5,39
Uranus,	0,97	Earth,	5,66
Saturn,	0,68	Venus,	5,22
Jupiter,	1,36	Mercury,	19,56

Sun, 1,47.

The primitive sphere from which these planets were successively expelled must in its early constitution have had its specific density determined through and through by the principle of gravity, making the superficial density proportioned to the central in the ratio inversely as the squares of the distances. When the planets were thrown off in succession, though circumstances may have somewhat modified this primitive ratio of densities, yet it is not probable that any great changes were wrought in the ratios of the densities from time to time as the planets were expelled from the diminishing circumferences. If they had condensed and hardened equally, there would doubtless have been perpetually a very exact uniformity in the ratio of their specific densities. But the difference of chemical elementary forces and their ultimate combination, incident to the positions in the primitive sphere and the times of expulsion, and the different gravities and radiations of heat and light under which the successive planets respectively cooled and settled down to their present volumes, must have induced considerable disturbances and derangements in the appropriate ratios of specific density. We need not wonder, therefore, when we find Saturn a little less dense than either Uranus or Neptune, and Venus a little less dense than the Earth, and Mercury so rapidly gaining in density upon Venus. The general augmentation is determined in the great principle of the ratio of gravity in the primitive sphere, and the interruptions in the uniformity have been occasioned by the interference of the contingent modifying influences. The control of the determining principle of gravity in the primitive sphere over the specific densities of the planets has been as constant and full as should have been anticipated.

The greatest departure from uniformity in augmenting density is found in the sun itself. As the great central body, its position should make its matter specifically the most dense of all, but in fact it is found far less dense than Mercury, and only about the consistency of Jupiter, as ordinarily estimated. Here is so wide a departure from the proposed determining principle, that if we could not find any corrections for the common calculation of the sun's density, and good reasons for this comparatively diminished solidity in its matter, we should be obliged to leave the undue rarity of the sun's substance as an utter anomaly. But we have already found from the openings in the solar spots that the sun has a luminous atmosphere of several thousand miles' thickness, and if this deep, imponderable superficies were subtracted from the received volume of the sun before the division of the mass by it, it would make a very great augmentation of determined density. Add to this the certainty, that both from its immense bulk and from its deep surroundings by an atmosphere of heat and light, the inner fires of the sun's body must be more intense, and the cooling process, if progressing at all, must be far slower than that of any planet, and the specific density of the sun will thereby be brought fully up to the point that the great determining principle of gravity in the primitive sphere would demand for it.

The generally augmenting densities of the worlds in our planetary system is as the principle of gravity necessitates, and the partial interruptions are the necessary results from the apprehended interfering circumstances.

There is, moreover, a very marked gradation in the *interplanetary spaces*, making a pretty regular diminution in breadth from the superior space between Neptune and Uranus down to that between Venus and Mercury, or that between Mercury and the sun. These distances determined by the distances of the respective planets from the sun between which they occur may be found as follows—estimated in round millions of miles:

			Interplanetary spaces.
Neptune,	2854		
Uranus,	1822	=	1032
Saturn,	906	=	916
Jupiter,	494	=	412
Juno,	254	=	240
Mars,	144	=	110
Earth,	95	=	49
Venus,	68	=	27
Mercury,	37	=	31

This marked gradation, very nearly duplicate in most cases, was very early observed, and long since different arrangements and appliances have been tried, to make the approach towards duplicate ratios of the interplanetary spaces to take on a more exact proportion. Kepler knew only of the 7 planets from the sun to Saturn, one of which was lost, or wanting between Mars and Jupiter; and he tried to find some mathematical explanation for so regular an augmentation of their distances by interposing different regular solids, or imagining the intervals of the musical scale between the planets. Bode, of Berlin, assumed the distance of Saturn to be divided into 100 parts, and the distance of Mercury to contain 4 such parts, and then assigning to Venus 3 parts and doubling the ratio for the planetary spaces successively beyond, he had contrived a formula that very well fitted the facts, and hence came to be known as Bode's law of planetary distances. Thus Mercury was $\frac{4}{100}$, Venus $4+3=\frac{7}{100}$, The Earth $4+6=\frac{10}{100}$, Mars $4+12=\frac{16}{100}$, small planets $4+24=\frac{28}{100}$, Jupiter $4+48=\frac{52}{100}$, Saturn $4+96=\frac{100}{100}$.

Wurm pushed the contrivance further, and by a more complicated arrangement of parts brought it still nearer to the facts. He assumed 387 parts for Mercury, and 293 from Mercury to Venus, and then doubled for each successive receding planet, giving the following formula—

		Attained parts.	Actual parts.
Mercury,	387	387	387
Venus,	387+	293=680	723
The Earth,	387+	2×293=973	1000
Mars,	387+	4×293=1559	1523
Juno,	387+	8×293=2731	2668
Jupiter,	387+	16×293=5075	5002
Saturn,	387+	32×293=9763	9538
Uranus,	387+	64×293=19139	19182
Neptune,	387+	128×293=37891	30036

Such contrivances are of no value, and if made exact could be of no use in attaining any philosophical explanation or determined necessity for such an order, and the only purpose subserved in here referring to them is, to evince how notable has been the fact of this gradation in the planetary distances. The widest departure from the duplicate ratio is in the last space between Uranus and Neptune, which is too small for the due proportion by nearly 8,000 parts, or if put into miles according to the fact, the deficiency is more than 800,000,000 of miles. This large deficiency in miles is, however, from the great distance of Neptune, only a deficiency of between $\frac{1}{4}$ and $\frac{1}{5}$ of the proper ratio. The fact of the nearly duplicate ratios in the planetary distances, manifestly indicates some generally determining principle, but this principle interrupted in its control by some interfering circumstances.

The general principle is in the gravity and revolution of the primal sphere of the system. The extreme planets, or those first thrown off, must have been the rarest, and their condensation must have left the largest spaces, the inferior planets growing more dense and leaving smaller spaces; and the gaining of the excess of revolution, after ejecting one planet so as to throw off the next, must have been as the diminishing circumference of the sphere, giving the shorter spacial interval to the smaller circumference. If Neptune be an outside planet, there must have been an outside friction that would prematurely separate it, and might thus account for its diminished interplanetary space.

The ratio in the *times of revolution* to the distance from the centre, is invariable in all the revolving bodies of the system. The so-called "harmonic law" prevails in all the planets and their satellites. The cubes of the distances are as the squares of the times of revolution. All the planets were therefore thrown off from the surface of a sphere, and not from the end of a line or the periphery of a circle. And since both the tangential impulse and the gravitating force remain constant after the ejection of the planet, undisturbed by any interfering influences as in the case of planetary densities and planetary distances, so the facts have here the exactness and universality of an unhindered determining principle, and therefore an invariable and unbroken law. That the satellites revolve about their primaries in the same ratio of the cube of the distance and the square of the periodic time as a matter of fact, has been determined in this, that the satellites were ejected from their rotating primaries in their ensphered state, and thus necessitating the cubes of the distances, and not the mere sum of the units, as would have been necessary in the ejection from the end of a line, or the square of the distances if the ejection had been from the circumference of a circle. The same facts in the solar, terrene, jovian, saturnian and uranian systems, have their necessary law in the third principle of planetary revolution.

The satellites revolve but *do not rotate.* An excess of rotation over adhesion or gravity, may very well occur to so great a degree in the original revolving mass as to give to some of the planets a force of ejection that shall secure their rotation not only, but so as also to throw off one or more satellites; but it can hardly be anticipated that the satellites shall have such a force of ejection as to secure their rotation, and much less that a rotation shall be sufficiently rapid to throw off sub-satellites. And the facts are, that not only is no satellite found with a sub-satellite, but no satellite is found to rotate on its own axis in its revolution about its primary. The moon, it is well known, keeps the same face to the earth through all revolutions, with the exception of certain slight librations clearly conditioned by two or three separate considerations. As the principle determines, the matter of the moon was just thrown beyond and separate from the matter of the planet, and while the revolution and rotation of the planet were conditioned by the tangential force of revolution in the great spherical mass, the revolution of the moon around the earth was conditioned by the force which was in the planets' rotation, and which was only sufficient to overcome the cohesion, but not to secure rotation. The moon, thus barely separated, keeps on its rate of revolution as if it had remained still in the circumference of the earth's mass of matter, as it was just preceding its first disruption, and thus perpetually revolves about the earth at that constant rate, while the matter of the earth rolls and condenses itself to a ball of continually more rapid rotations beneath it. The moon, thus, having no rotation, and therefore no balancing upon its own axis, cannot revolve as the earth in its rotations does, with an axis always parallel to its former positions in the orbit, but with its centre bound to the earth's centre as if it were still held by the radius of its old first revolution. It thus keeps constantly the same hemisphere to the earth, and as if its old radius was a cord attached at opposite ends to the centres of the two bodies, the moon's one face thereby necessarily turns to every portion of the earth's surface with each revolution.

The same conformity with the principle is found in the facts of the very *slight eccentricity* of the moon's orbit, and the *absence of all flattening at the poles.* If the moon had been ejected from its primary with sufficient force to rotate, it must have been considerably elliptical in its orbit; and if it had rotated on its axis, it must have been oblate proportioned to the rapidity of rotation. The facts all correspond to the determinations of the rational principle.

And not only with the moon, but the same accordance of principle and law is found also with the *satellites of all the planets*, so far as any discoveries of facts have been made. Sir William Herschel has found satisfactory indications, in his most accurate observations, that the moons of Jupiter have only the same enlightened hemispheres to the planet; and the largest satellite of Saturn is also found to have a variation of apparent brightness in different parts of its orbit, and that the same brightness always corresponds to the same position on the surface of Saturn in the revolutions of the satellite about it. This general law of the satellites, that they constantly turn one face to their primary, has been sometimes accounted for by supposing that one hemisphere of the satellite is protruded towards its planet, and thus held in place by an excess of gravity in the protruding part; but no fact of such protuberance appears, and the true principle determines the facts as they are given, without any such gratuitous hypothesis.

So, moreover, in the *planes of the orbits*, and the *direction of revolution*, we have the same remarkable accordance of fact and determining principle. The great revolving sphere of which the system is to be constituted, must throw off its planets and satellites successively, and the great central force will keep this wheeling sphere revolving in one general direction through all the process. The outermost or earliest planet will be thrown off in the same general direction as the innermost and last formed, and thus the courses in their orbits will be direct, and not retrograde. At the same time the modifications of the central current will very probably secure some oscillation of the wheeling sphere on its own centre, and thus the equatorial plane must have some changes of direction. This would necessitate corresponding varieties in the planes of the planetary orbits, and while all cannot be expected to be formed in the same plane, the general uniformity of the revolving-force will not admit of very wide varieties.

The facts are, that all the planets move in one direction in their orbits, or from west to east as viewed relatively to the terrestrial axis, and all the planes of the planetary orbits are inclined somewhat to each other, but still within very limited degrees. No two are parallel, nor exactly in the same plane, and the widest extremes, aside from the planetoids, do not vary but about 7 degrees. Taking the sun's equator as the present fixed plane, we shall find a pretty uniform oscillation from planet to planet, till we get to Mercury, and which suddenly drops into very near conformity with the plane of the sun's revolution. The oscillations are little removed from one degree between the successive planets till we come to the earth, which between Mars and Venus oscillates from two to three degrees. Thus—

Neptune, ...	5° 43′ 00″
Uranus, ...	6° 43′ 32″
Saturn, ...	5° 00′ 25″
Jupiter, ...	6° 11′ 09″
Mars, ...	5° 38′ 54″

Earth, ...	7° 30′ 00″
Venus, ...	4° 06′ 32″
Mercury, ...	0° 29′ 55″
Sun, ...	0° 00′ 00″

But when we take the planes of the orbits of the satellites, we have, as we might anticipate, much wider extremes; and both in the extremes of orbital planes, and more especially of direction of revolution, we have really an astonishing conformity of fact and principle. The great wheeling sphere of matter must revolve on pretty uniformly, and the slight oscillations can occasion but little inclinations of the planes of the planetary orbits, and no retrogradations apparent in the course of the planetary revolutions. But when the planet is ejected, many opportunities for modifying causes occur. The rotation of the planetary matter must conform in its axis, to the composition of the ejecting force which throws the matter forward and the attracting force of gravity bringing the matter backward, and thus wrapping the upper portion over the lower; and according to the equal density of the matter, and the directness of the ejecting force, must be the regularity of the axis of rotation. Exact equality of density, and precise projection of revolving force, would secure the rotary axis of the planet directly parallel to the axis of the principal sphere, but any inequality of density in the matter to be thrown off at the circumference must give to the projecting force a modified direction, and turn the impulse, and thus the direction of the ejected matter, more from one side of the equatorial plane than from the other. If the denser matter be on one side of the equatorial plane, the tendency will be to throw the matter towards the other side, and the axis of rotation must be directed accordingly; and thus it may be, that while the general tendency is toward an axis across and perpendicular to the plane of the planetary orbit, specific cases may give the axis of rotation nearly in the plane of the orbit. A very dense lump on one side of the ejected mass might turn the axis of rotation to be almost like the rifle ball, or near 90° from its regular direction perpendicular to the plane of the orbit. The rotation must, therefore, be in the direction of the projecting force, and necessarily direct and not retrograde, and yet the axis of rotation in one planet may very well so be turned in inclination towards that of another planet, that the satellites of one may appear to have a retrograde movement when viewed from the other. And so the facts really are found to be in our own solar system.

Mercury is too near the sun, and too perpetually within its strong light, to determine its rotation by any observation. Venus can be known to rotate from the different appearances of the cusps or horns while passing through its illuminated changes, but the direction of its axis has not yet been, and perhaps may never be, determined. So far as we have the determination of the direction of the planetary axes, either by direct calculation, or by deduction from the general plane of the orbits of their satellites, they may be given as follows, in the degrées of their inclination to their own orbits respectively:—

The Earth,	66° 32′
Mars,	61° 18′
Jupiter,	86° 54′
Saturn,	115° 41′
Uranus,	168° 12′

Leaving Mars, which has no satellite, and estimating the above in reference to the ecliptic or the earth's orbit, and also in reference to the earth's axis, we shall have the following:—

	Inclination of axes to the ecliptic.	Inclination of axes to the earth's axis.
The Earth,	66° 32′	00° 00′
Jupiter,	88° 13′	21° 41′
Saturn,	118° 11′	51° 39′
Uranus,	168° 58′	102° 26′

If we consider the orbits of the satellites as perpendicular to the axes of their primaries, which must doubtless be very near the fact, that of the moon inclining a little more than 1° 30′ to the axis of the earth, we should have the following appearances of the satellites of the different planets from the earth, as determined by the principle of the necessary revolutions and rotations of the respective planets. All these planets must move in their orbits in the same general direction as the earth, for they are successively thrown off from the same spherical mass; and this must also secure that, in reference to their own orbital movements, their rotary movements must also be in the same way direct, and not retrograde, for their rotary movement must be the result of the projectile force which separates them from the wheeling sphere, and though unequal densities in the matter may greatly modify the inclinations of the axes, yet must the rotations on the axes in all cases be before and not against the projectile force. In the case of the earth, we find the moon moving from west to east in her revolution about the earth, and thus determining that the projectile force of the wheeling sphere was in that general direction, but with such an inequality in the density of the matter as to turn the axis 66° 32′ out of its regular perpendicular position to the earth's orbit. With this direction of the earth's axis for our north point, and the rotation of the earth, and thus the revolution of the moon from west to east, and making the earth our stand-point, we must find Jupiter's moons also moving from west to east, and in orbits inclined to the earth's axis of generally 68° 19′, thus giving only a profile view, somewhat narrow, viz., 21° 41′, to the Jovian system, from our earth. The Saturnian system must come much more broadly in profile, for its inclination of the orbits of its moons generally must be 38° 21′, giving a face of 51° 39′ to the view from the earth, and the moons still moving from a western to an eastern direction, for their orbits are still above or west of the north polar point of the earth's axis.

But when we come to the Uranian system, we have the orbits of its moons inclined 12° 26′ to the earth's axis from the under or eastern side, so that the face has really turned itself by the full plane, and has cast itself a little in profile on the other side, viz. 77° 34′. The moons of Uranus, as seen from the earth, though really moving in the general direction of the planet itself in its orbit, and thus truly direct and not retrograde, must yet appear from the earth to be moving from an easterly to a westerly direction. The axis of Uranus is 11° 48′ turned across the plane of its orbit, and must therefore rotate in the general direction of the projectile force from its orbit, but because the axis of the earth is more than 90 degrees turned from the axis of Uranus, the moons of Uranus must from the earth appear to move in a westerly direction.

That the moons of Uranus are retrograde has been a surprising anomaly from its first discovery, but that this exceptional fact is found to leap within the necessary determinations of the eternal principle, and is found anomalous only in appearance, the principle itself expounding why it must so appear, is a most conclusive example of that accordance of fact and principle, which is alone true science.

The marked peculiarities of the *Planetoids* have been matters of great interest and wonder from the earliest discovery of any of their number. The wide breach, in the pretty nearly duplicate ratio of increase in the interplanetary spaces from the centre outwards, which occurs between Mars and Jupiter, had been very extensively considered by astronomers as the place of a lost, or of as yet an undiscovered planet, and at the close of the last century much interest had been excited to extend a careful observation over all this region, with the hope of finding such missing planet. On the 1st of January, 1801, the Italian Astronomer Piazzi, discovered a star that seemed by the next evening to have changed its place, and by repeated observations during the month, he had determined its truly planetary character, and though lost in the sun's light, it had again been found after emerging on the other side of its orbit, and was determined in its elements and called Ceres. It held almost exactly the right position for dividing properly the space between Mars and Jupiter, but was exceedingly minute as a planet, its diameter being only about 160 miles. In March of the following year, Dr. Olbers discovered Pallas, having nearly the same mean distance as Ceres, but with an orbit much further inclined from the ecliptic. In September, 1804, Juno was discovered, and in March, 1807, the discovery of Vesta followed, after which no more planetoids were found until 1845, when M. Hencke discovered Hebe, and since that time the search has been so diligently and successfully pursued, that in 1856, forty planetoids had been found and their elements calculated.

These small bodies all revolve within a limited space between Mars and Jupiter, and may very well be deemed as filling the appropriate place of one planet. Mars is about 145,000,000 of miles and Jupiter about 494,000,000 of miles from the sun, and these planetoids range between 209,000,000 miles and 300,000,000 miles in mean distances from the sun, thus occupying about 90,000,000 of the 349,000,000 miles of this interplanetary space, and in the proper ratio nearer to Mars to conform to the general ratio of the other interplanetary distances. The largest of these bodies cannot be more than 500 miles in diameter, and the smallest may not be 50 miles, so that the aggregate of all the asteroids yet discovered, would make but a small planetary body if united. While thus occupying the space due to one planet, and together conforming to the general conditions of the planetary bodies, yet are there some pretty wide differences among themselves, and some marked peculiarities from other planets.

Their orbits, though elliptical, yet widely differ in amount of eccentricity from each other, and the least eccentric is still more elliptical than any of the planets, except Mercury and Mars, while the most elliptical about doubles the eccentricity of Mercury. Their inclination to the ecliptic, in their orbits, is also of a much wider range than in the case of the planets. They all lie on one side of the earth's orbit towards the sun, but there are 23 between the ecliptic and the plane of the sun's equator, and 17 beyond the plane of the sun's equator. The nearest to the ecliptic is Massalia, 0° 50′ 16″, and the furthest inclined from the ecliptic is Pallas, 34° 37′ 20″. Their longitudes of ascending nodes, or the lines in which their orbital planes cut the plane of the ecliptic, are widely extended; Fides being about 8° and Atalanta about 359°. Their difference in longitude of perihelion puts also their directions of their major axes all around the sun, the least being that of Lœtitia about 1°, and the largest being that of Polymnia about 341°. Their movements in their orbits conform to the planetary principle that the cube of the mean distance is as the square of the periodic time, and thus the most distant has the least velocity in its orbit. Flora is the least distant from the sun, whose semi-axis of orbit is to that of the earth as 2,2017, and its daily motion 1086″, and its sidereal period 1193 days; while Euphrosyne has a comparative semi-axis of orbit with the earth of 3,156, and a daily movement of 633″, and a sidereal period of 2048 days. Their movements are all direct in their orbits, and all their general conformities with the planets are determined in the same principle as that which gives to them their universal law. But their peculiarities, so widely distinguishing them from all the regular planets, have hitherto been brought under no determining principle, and subjected to no known laws. Their small masses, their widely extended degrees of inclination, longitudes of ascending node, and longitudes of perihelion, with their great differences of eccentricity. are considered as yet wholly anomalous.

But the immediate circumstances in the formation of the planetoids, when fully considered, will fairly bring all their peculiarities under the determination of the one great principle that has ruled in all the planetary constitutions. When the great planet Jupiter, whose mass is more than 338 times that of the earth, had been just thrown off, its attraction upon the wheeling sphere beneath, and from which it had just parted, necessarily put that sphere in a peculiar position for the formation of its next equatorial accumulations. Without here considering how the action of Saturn would necessarily, if in conjunction with that of Uranus, tend to the ejection of so large a mass as Jupiter, the insight into the action of such a mass as Jupiter gives just the conditions necessary for forming the planetoids. As the primal sphere was revolving under so large an attracting body as Jupiter, which yet might not have condensed much within the orbit of its outermost satellite, this sphere must have had its equatorial accumulations thereby hastened, and the general equatorial protuberance induced by the revolution and hastened by Jupiter's attraction must also by this attraction have been considerably disturbed, and drawn from an equable equatorial diffusion over the surface to a rising tide directly under this large planet. The same attraction that made the tide on the side next to Jupiter must also have lightened the antagonism at the centre of the sphere, and thus just balanced the pressure upon the opposite hemisphere, and therefore raised simultaneously an equal tide upon the side of the sphere opposite to Jupiter.

In such a state, the rotating sphere could not hold on and retain its equatorial accumulations, until the tangential force should bring all into the equatorial ring and expel all in a mass, but as the accumulation had become considerable in the equatorial region, and Jupiter's attraction had brought up a tide beneath, so the tangential force must first have caught the crest of that tidal wave and expelled it from the main sphere, and Euphrosyne the outermost planetoid was thus driven upon its separate revolution. The same attraction above, and the same tidal wave and revolution beneath, anon threw off Hygeia, and then Themis, and Leucothea, and so onward in successive instalments till we reach Flora, the last and least distant from the sun of any that has been yet found, and the balancing relief was gained as if all had been expelled in one planet, until the ordinary accumulation again returned with no progressing attraction for a flowing tidal wave, and Mars was made a regular though smaller planet. The accumulation and tide must have made the first planetoid exclusion untimely early, and the last untimely late, and hence the width of the planetoid orbital region, and hence also the disproportioned smallness of Mars.

As Jupiter passed on his orbit, and the primal sphere rotated beneath, the crest of the tide must have been perpetually varying through all the equatorial circle, and hence the planetoids must have gone off at all their different degrees of perihelion longitude and ascending nodal line, and as there must have been unequal attractions at different times on opposite sides of the equatorial circle, so there must have been all the different degrees of inclination that the planetoidal orbits possessed, and which may perpetually be modified by later interferences. With such a neighboring planet as Jupiter, its next inferior body could not have been thrown off a mature planet, but must have been a planetoid followed by others in succession till the accumulations and expulsions were balanced. And on the other hand, nothing but such preponderance of planetary attraction and tidal elevation progressing from place to place over the equatorial circle, could have combined so many peculiarities of planetoid formation and revolution as the facts disclose. The planetoids must have been between Mars and Jupiter; they could not have been produced between any other two of our planets.

Of precisely an opposite character to the planetoids are *the rings of Saturn.* Many of the planets have satellites, but Saturn only is surrounded nearly in its equatorial plane by a system of concentric thin and broad rings. So singular a fact has rendered the phenomenon a matter of the highest interest, but no investigation has as yet brought the fact under the determination of any principle, nor subjected it to any necessary law. Saturn has eight satellites, all but the exterior one in the same general plane of the ring which is interior to them all. The exterior satellite is inclined to this plane 12° 14′, and is more than 64 semi-diameters of Saturn distant from it, while the interior satellite is but a little more than 3 semi-diameters, or about 144,000 miles, distant from the centre of Saturn. The ring is manifestly, at least in some periods of its observation, divided into two, one within the other, with a comparatively narrow space between them. The exterior ring is the narrowest, and is 10,573 miles broad, the interior ring being 17,176 miles in breadth, and the interval between the rings being 1,791 miles in breadth, thus giving to the whole annular space a breadth of 29,540 miles. The interval between the lower edge of the ring and the equatorial surface of Saturn is 19,089 miles, or about one-half the semi-diameter of Saturn. There has recently been discovered a third transparent ring, stretching down from the inner portion of the second ring like a thin vail toward the surface of the planet. The thickness of the ring is so small that when its edge is alone presented to the best glasses it is invisible, or barely perceptible, and cannot be more than 250 miles. It is slightly eccentric, and thus balancing itself upon a movable point about the centre of Saturn, and is deemed to be in a fluid state.

If now, when the revolutions of Saturn had brought the equatorial accumulation to a pretty equable and narrow distribution about the mid-line of the planet, and this raised equatorial ring was passing round in regular rotation with the planet, and the tangential force hardly availing to throw it off as it had done its last satellite, the eight moons of Saturn should be so distributed in their revolutions as to lift, in their attractions, pretty equably on all sides of the planet around its equator, it must follow that the body of the planet would separate itself from this equatorial accumulation, and the separated part would at once be a fluid ring henceforth perpetually to revolve at its own steady velocity, and in its own place, about the body of the planet as that should condense and rotate beneath it. The rotating planet, and the revolving ring, together with the revolving moons, would all be carried about the sun, by the tangential force that sent the planet out in its orbit.

This ring, thus separated and revolving, must throw its

fluid matter upward and spread itself out into a thin plane, and any subsequent unequal attractions must make partial divisions, and continued condensations would secure complete separations all round within the substance of the ring, and would thus necessarily work it into the precise position presented by its present phenomena. Nothing hinders that this ring may perpetuate itself unbroken transversely, so long as its fluid state permits it to yield and equilibrate itself to any disturbing one-sided attractions, and its rotating centre about the planet's centre to balance all disturbing revolutions.

Nothing could give such a revolving ring, but such a favorable distribution of many attracting bodies at the right state of the equatorial accumulation; and if such an arrangement of circumstances did occur, then Saturn's ring was a necessary consequence. No other planet has the requisite satellites, and no other occasion could be given in our system for a revolving ring, but in some such distribution of planets about the primal sphere, and that indeed may have made the ring between the Earth and Mars that reflects the mysterious zodiacal light upon us.

11. The Law of Comets.—Remnants of the chemically chaotic matter, which have not been taken up and condensed in the world-systems, must come together in the diremptive currents that yet move through the primitive ether, and thus form larger or smaller collections of matter moving amid the intervening spaces of the systems. They may also be constantly accumulating in the irradiations and interworkings of the primitive forces, and thus at no time will the interstellary spaces be free from many light and floating bodies, that condense about their own

centres with sufficient consistency to maintain their integrity of connection, though rapidly moving through the ethereal fluid. These bodies are comets, and subject to the universal forces that constitute and move them, but as independent of the systems they manifest themselves to us by no phenomena, and give no facts to be referred to the control of any particular laws. They may come within a system, pass through and beyond it, and have no more any communication with it, or be caught and retained by it, and afterwards make up a component part of it, and have their facts bound up in laws that are necessitated by the principles inherent in it. Under this idea of cometary origin, we contemplate the facts which the comets connected or communicating with our system give to our observation.

Comets manifest themselves to be exceedingly tenuous in the matter of their composition, inasmuch as they are completely diaphanous, and the stars before which they pass appear through their most central portions with their light not perceptibly diminished. But this extreme rarity compared with the bodies of the planets is still great density compared with the primitive ethereal fluid. They pass, sometimes with the most surprising velocity, through this ethereal medium with but slight obstruction from it. The ethereal resistance, wholly inappreciable by any reference to the movements of the planets, is yet quite determinate in the movements of the comets. Encke's comet has lost distinctively in each of the last ten observed revolutions, what in the aggregate amounts to about one entire day. There has also been determined in the revolutions of Biela's comet a similar retardation. So also, when the tail of a comet appears curved, the convex portion is found to be in the direction from which the comet has been moving, therein manifesting that the most subtle perceptible vapor of a comet is still impeded by the ether through which it passes. Such thin and light bodies cannot have sufficient gravity to induce a radiation of any latent heat-force at the centre and make them to be self-luminous, and hence we find they do not, like the stars, keep the same brightness, though diminishing in apparent volume by distance, but often fade out and disappear even where the body at the last still presents a considerable disk, therein manifesting that they have shone only in reflected light.

The principle of cometary origin admits that they may be deflected from their former course by the interfering attractions of a system, and made to take a perihelion passage about the central body, and thence may pass on in a curve that returns again completely within itself, and become an elliptical orbit, or it may pass on and pass out of the system in either a parabolic or hyperbolic curve, and never again return to the system through which it has passed. A comet, also, that has been caught and retained in a complete elliptical orbit, may subsequently be subjected to interfering attractions that shall greatly change its orbit, or even make it to take a curve that shall completely carry it again out of the system. The retained comets will have more varied inclinations and eccentricities than the planets, for they come in under impulses of all degrees and directions, while the planets are ejected from the one source of the same rotating sphere. Some will pass about the sun in one direction, and others in an opposite direction, and thus there will be comets of direct and others of retrograde movement, the chances for each being equal.

The facts correspond to what might thus be anticipated. Faye's comet, discovered in November, 1843, was determined to a period of 7½ years in an elliptical orbit of but a little extent beyond Jupiter. The wonder was, why a comet so often occurring should not have been before observed? The answer which has satisfied, was derived from the fact, that in the point of its aphelion it had then approached very near to Jupiter in that part of his orbit, and that thus it had been deflected from its course in a parabolic curve, or a much larger ellipse, and was thereby then brought into an orbit within the range of terrestrial observation.

A more remarkable case to the same point, is the comet discovered by Messier in 1770. This comet made two revolutions round the sun, and has since been lost to further observation. Very careful calculations have been made by Lexell, La Place, and especially by La Verrier, and with somewhat different conclusions, but all concurring in this, that the disturbing influence of Jupiter in its last aphelion had thrown it quite out of its former orbit, and that it might have gone out wholly from the system, or into another elliptical orbit not yet recognized, or might even have passed within the orbits of the moons of Jupiter, and been permanently lodged in that planet. The general fact seems established, that comets have been both caught in, and expelled from the system.

Biela's comet has its orbit from a little within the earth's orbit to a little beyond Jupiter's orbit, and thus of course moves almost entirely between the earth and

Jupiter, passing the orbits of Mars and the planetoids. It is but little inclined to the plane of the general system, of comparative slight eccentricity, and in a period of less than 7 years revolution. These planets may act upon it with a divellent force in opposite directions, but their joint action in conjunction must be too slight, from the necessary distance in such a case, to be of any account. The dense bodies of Mars and the planetoids may also interfere occasionally with great effect.

The very singular fact that this comet separated itself into two distinct comets in its appearance in 1846, and that so long as they remained visible for about 4 months, they continued separate at rather diverging directions, and especially that on their return in 1852, they had increased their direct distance apart to more than one and a half million of miles, is a clear evidence of the slight consistency in the bodies of the comets and the readiness with which interfering influences may greatly modify their entire constitutions. Whether conflicting attractions or disparting resistances induced this disruption, the fact itself determines that comets may readily be multiplied or changed.

Up to 1854, of more than 200 comets that had appeared and been determined in their elements, we have the following distinctions with their peculiarities:

(1.) Comets with elliptical orbits. There are 13 of this class whose mean distances are within the orbit of Saturn. Taking the earth's mean distance as the standard, the least distant of these is 2,2148, and the most distant is 6,3206. The least eccentric of them is 0,6173, and the most eccentric is 0,8490. The orbit least inclined to the ecliptic is

1° 34′ 28″, and that most inclined is 30° 57′ 51.″ They all move direct in their orbits, or in concurrence with the planets. There are again 6 of these whose mean distance is within and pretty near to the orbit of Uranus. The least distant of these compared with the earth from the sun, is 14,5306, and the most distant is 17,9875. The least eccentric is 0,9248, and the most eccentric is 0,9726. The least inclination to the ecliptic is 17° 45′ 5″, and the most is 84° 57′ 13″. One of these, Halley's comet, is retrograde, and the others are direct. And then, further, there are 21 of the comets with elliptical orbits whose mean distances exceed the furthest known limits of our system. Compared with the earth's solar distance, the least is 33,0310, and the furthest is 2138,0000. The least eccentric is 0,96990, and the most eccentric is 0,99998. The least inclined to the ecliptic is 21° 16′ 5″, and the most inclined is 83° 47′ 46″. Ten are direct and eleven retrograde. There were thus determined 40 of the elliptical comets.

(2.) Comets with hyperbolic courses. There have been determined in their courses 7 hyperbolic comets. The perihelion distance of the least, compared with that of the earth, was 0,6184, and the furthest was 4,0635. The least inclined to the ecliptic was 11° 15′ 19″, and the most inclined was 83° 20′ 26″. One was retrograde and six direct.

(3.) Comets with parabolic courses. There have been determined 160 parabolic comets. The least in perihelion distance must have nearly grazed the surface of the sun, and the furthest was in perihelion distance, compared with the earth, 2,1985. The least inclined was 1° 55′ 0″, and the most inclined to the ecliptic was 89° 22′ 10″ Of these 70 were direct, 86 retrograde, and 4 not determined.

Some of these statistics distinguishing the comets from the planets are, their wide differences of inclination from a little more than 1° to a little less than 90°, while the inclination of the planets only reaches to about 7° and the planetoids to about 35°; and also their degrees of eccentricity in their orbits, which compared with the earth is from 0,6173 to 0,99998, while the most eccentric of the planets is Mercury, 0,38709, and of the planetoids is Polymnia, 0,337. Their mean distances from the sun are from a little more than twice that of the earth to 2138 times the earth's distance. Such wide discrepancies determine that the planets and elliptical comets could not have been thrown off from the same rotating sphere, and that the comets could not themselves have been formed from any one systematic revolution, but must have come within the system under separate impulses. This is still more manifest in the wide differences of longitude of ascending node in the comets, reaching from 1° 12′ 24″ to 356° 17′ 38″. The most elliptical planetary orbit is still near to a circle, but the most elliptical cometary orbit has the two sides and the major axis almost parallel to each other. Such wide extremes are plainly incidental to their separate and independent introduction into the system.

But while these general facts come thus under the general law for cometary connection with the planetary system, there are two particulars that require and will repay a distinct and close examination. We will *first* look at the facts of the inclination of cometary elliptical orbits to the ecliptic; and *second*, to the facts connected with their direct and retrograde revolutions.

We have seen that of 40 elliptical orbits of the comets, 13 are within the orbit of Saturn, taking the mean distance, and their extremes of inclination to the ecliptic are from 1° 34′ 28″ to 30° 57′ 51″; and that there are 6 within the orbit of Uranus and near to it, whose extremes of inclination to the ecliptic are 17° 45′ 5″ up to 84° 57′ 13″; and there are 21 beyond the outermost known planetary orbit, ranging in extremes of ecliptic inclination from 21° 16′ 5″ to 83° 47′ 46″. The peculiarity observable is in the first class of orbits within Saturn, where their extremes of inclination are quite limited, the furthest being still within the plane of the furthest inclined planetoid, and all the rest within 18° of inclination. In the other two classes the extreme inclination reaches almost to a direct perpendicular to the ecliptic. Moreover, in the first class the orbits are much less eccentric and their elements strongly allied to those of the planets, so that they have been distinguished as having a planetary character and especially as strongly analogous to the orbits of the planetoids. The extreme rarity of all comets will, however, universally distinguish them from all planetary bodies that have been thrown off from their one primal rotating sphere. The nearest as well as the most remote in their superior apsides, all have come within the system from some independent source without. But why the nearest class so conformed in inclination to the planetary system, and the others reach almost at right-angles to it?

Entering the system independently, they should have had each an equal ratio of extreme inclination on their first introduction, and hence we must look for something within the system that has induced this peculiarity since their entrance. Now a careful insight into their position and revolution in reference to the planets will disclose this necessary law of their limited inclination to the ecliptic. Suppose one of these comets on its solitary way to have come within the gravitating force of our system, and to be so drawn to it that it passes in and around the sun, and onward in its elliptical path to a complete orbit, and that the plane of this orbit is directly perpendicular to the common plane of the planetary orbits. In its revolution it does not go off further from the sun than Saturn, and though at right-angles to Saturn's course it will not at the furthest remove have gone beyond all Saturn's appreciable influence, and through all its revolution it will be affected by all the planets within the orbit of Saturn. These planetary attractions will not always, nor even often, be all uniformly arranged so as to balance their action upon the comet, but must almost perpetually act in excess upon one side of the comet, and which must turn it down and incline its orbit less than at right angles to the planetary plane. Such inclination within 90° once secured, the conspiring attractions of the planets in their perpetually concurring occasions must bring the inclination nearer and nearer into conformity with the planets, until the whole cometary orbit shall find its place of general equilibration with the attractions to which from perihelion to aphelion it is subject, and must afterwards oscillate about this as alternate excesses and deficiencies of gravitation on each side shall induce. Any comet, commencing to revolve in an orbit less than a right-angle to the common planetary plane, must the more certainly be at length brought to its equilibrating plane of inclination, and as these attractions are from the bodies of the system, so they must bring the cometary orbit into a general conformity with the planetary orbits.

On the other hand, when the length of the orbit puts the comet through the superior half of its orbital movement beyond the reach of any appreciable planetary attractions, the general plane cannot be modified thereby, and it must perpetually revolve at the same general inclination of the orbit as was first originated. The smaller orbits also must have had a less excess of projectile over the attractive force to carry them round the sun, and thus they must be less eccentric and more circular.

But a still more surprising conformity appears in the facts of direct and retrograde movement by the comets, to which we will next attend. All the comets whose mean distances are within Saturn are direct in their movements; all but one are direct also whose mean distances are nearly equal to that of Uranus; and of the 21 comets determined, whose mean distances are beyond the outskirts of the system, 10 are direct and 11 are retrograde. Why this growing tendency to retrogradation in the greater distance?

If we suppose a comet whose distance and plane of inclination brings it perpetually within the attraction of the planetary bodies, it will be found a natural and necessary result, that such comet shall ultimately take on a direct movement in its orbit. With such an inclination, no matter what its original longitude of ascending node, nor what its longitude of perihelion, the conspiring movements of the planets in their orbits and of the comet in its orbit will necessitate, that a comet of retrograde movement shall at length take on a direct movement. The planets all move in one direction; as estimated from our terrestrial standpoint, this direction is from a westerly to an easterly bearing. Suppose the comet, no matter what its longitude of perihelion, nor what its orbital plane of inclination if brought to lie beyond the sun's semi-diameter and yet not beyond the reach of all planetary attraction, to have such a longitude of ascending node for its orbit as not to equilibrate with all attractions through the whole line in which its plane cuts the plane of the ecliptic. Suppose this longitude of ascending node at first to be 45° from the point of Aries, and that the movement of the comet is here retrograde.

This comet must, then, in its revolution, have all the planetary attractions it can meet on the right hand side of its course, to come from planets that are moving in opposite directions to itself, and which must therefore be to and past itself; and all the planetary attractions it can meet on the left hand, to come from planets that are moving in the same direction and thus for some distance concurrent with itself. The aggregate amount of attractions received by this comet, in any considerable number of revolutions, must be the largest on its left hand side, and must therefore secure a turning of the orbit to the left and thereby a proportional elevation of the longitude of ascending node. It is true that incidentally to some revolutions, the conjunctions or oppositions of the planets may be such in reference to that particular revolution, that the attractions shall then happen to be the most on the right hand, and the longitude of ascending node be diminished instead of being increased; but yet, taking one revolution after another for successive periods, the attractions of the concurrent passages must exceed in the aggregate the attractions of the occurrent passages, and thus, with some oscillations, the line of ascending node must rise higher and higher from the point of Aries.

That such increase of longitude of ascending node may be very considerable in a single revolution is very manifest from the observed interference of planetary attractions with the regularity of cometary revolutions. Clairaut determined that Halley's comet, from perihelion in 1682 to next perihelion in 1750, had been disturbed by Saturn's attraction so much as to increase its period 100 days, and by Jupiter's attraction so much as to augment its period 518 days, thus making from both influences the comet's revolution to be more than 20 months longer time than its proper sidereal period. So great a retardation from planetary attraction in one revolution, and this moreover from two planets only, is sufficient to evince how readily the line of ascending node may be modified. Ultimately, our supposed comet whose orbit has a longitude of ascending node at first of 45° must pass on beyond 90°, and having thus passed the culminating point from Aries, the comet in its course instead of now running against the planetary movements on the right hand side of its major axis, will be running with the planetary movements on the left hand side of its major axis, and thereby have changed its relative course from a retrograde to a direct movement. The same excess of attractions on the concurrent side of the orbit with the planets must make the line of ascending node still revolve to a greater degree, till the point is reached in the particular orbital plane of the comet, that equilibrates the right and left hand attractions through the whole revolution, and must there remain with the slight oscillations to and fro that incidental disturbances will occasion.

Any other supposed degree of longitude of ascending node than 45°, must be subject to the same excess of attraction on the side of the concurrent passages of the planets, and bring the comet to a conformity of relative movement with the planets, and fix its orbit in its balanced position. Thus all comets that move in orbits wholly within the planetary system must at length become direct, even if primarily they were retrograde.

Now, all the comets whose mean distances are within the system are direct in their movement, except the one case of Halley's Comet. If that one case remained with no tendencies to come to a direct movement, it would throw its doubt upon any such principle as determining the law. But if that solitary exception clearly manifest that it is working under the control of this principle, it becomes itself a fact in confirmation, and we do not need to have had the observation of any past case, since we can see that the last and only remaining case is surely coming into conformity. The last 7 revolutions of Halley's Comet have been determined in their elements, and the changes of longitude of ascending node are as follows:

The orbit for perihelion	passage	1378	had lon. of	as. node	47° 17′ 00″
"	"	1456	"	"	48° 30′ 00″
"	"	1531	"	"	45° 30′ 00″
"	"	1607	"	"	48° 20′ 28″
"	"	1682	"	"	51° 11′ 18″
"	"	1759	"	"	53° 50′ 27″
"	"	1835	"	"	55° 09′ 59″

Here then is a steady and pretty equable increase for each revolution, but that of perihelion passage 1531, and which must have had an uncommon combination of opposing planetary passages, if the calculation has been correct. On the whole there is an increase of 8° in 7 revolutions of the comet, and though there be but the same increase in future, the orbit will pass the perpendicular and the comet be direct after about 30 more perihelion passages. That Halley's Comet is thus approaching to a direct movement is a significant fact for the principle that has determined all that may have been retrograde, long since to have become direct as they are now found.

Those comets whose mean distances are out of the system, and some of them more than 50 times the mean distance of Neptune, can have but little interference from planetary attraction in their rapid perihelion passages, and none that can bear upon the aphelion half of their orbit, and therefore no steady approaches to any change in the order of their orbital movement can be anticipated. These comets should perpetuate their movements either retrograde or direct, as their primitive introductory impulses have determined for them. The chances for each were equal, and the facts give as near an equal division as possible, for of the 21 comets revolving beyond our system as above determined, 10 are direct and 11 retrograde.

The impulse and attractions that make a hyperbolic cometary course must be extreme, and therefore few are formed. There were of the 7 determined hyperbolic courses, one comet only retrograde. A parabolic course is the most probable for a comet, for it most nearly equalizes the projectile and attracting forces, and excludes the incidental interferences that should turn it to an ellipse, and we find determined 160, of which 70 were direct, 86 retrograde, 4 not ascertained. As near an equal division as a promiscuous determination would lead us to expect.

12. Laws of Geological Formation.—The one Idea for the formation of worlds, by the revolutions of rotating masses of molten matter, applies to all stars and systems, but the facts can be brought within observation and experiment only on our globe, except as some very general phenomena may be discovered in the moon, and the other planets and satellites of the system. The earth also is completely hidden from our observation, except to the depth of some eight or ten miles of its superficial portion. But all the facts of Geology that we find conform to the conditions imposed by the determining principle we have attained.

The grand facts in the superficial crust of the earth are, that it has been broken up by some action of subterranean forces, and large masses of the broken strata have been made to turn up their edges to the surface with a greater or less dip toward the horizon, and these upturned edges exposed to view, disclose what are the orders and the contents of the several strata as they lie in their undisturbed horizontal position. The lowest depths disclosed determine unmistakably the near and constant action of intense heat. The great base of all the discovered and approachable strata is the *granite*, which has been cooled and crystallized, and lies upon the burning fluid beneath in a solid mass of unknown thickness. Above this lie the *gneiss* formations unstratified and of great thickness, and on this rests the superincumbent stratifications of *mica schist* of many thousand feet depth. Herein is embraced what some Geologists classify as the *Cumbrian formation*, and in which nothing but the chemical composition of unorganized matter appears. Thus far only the antagonist and diremptive forces have blended, and dead matter only with no traces of the life-force appears.

The *Cambrian* System of *old slate stone* a mile in thickness through its varied stratifications overtops the Cumbrian, and in those strata we begin to find the evidences of air and water, and that the slate beds in which the lowest fossil remains are found, must have been deposits beneath the wâter and not the cooling crust above the fire. The *Silurian* system is above this for a mile and a half in thickness, and its various stratified deposits have their hundreds of extinct species of fossil organizations; and then we have *the old red sandstone* many thousand feet thick, made up of the fractured and decomposed rocks which have been rent asunder and here deposited from some older formation, with many fossils of wholly extinct species. Then the interposed *limestone* and *coal* formations, the *new red sandstone* the *oolite* and the *chalk-beds*, all of several miles depth, finish what is known as the *secondary* formation. Higher up still is the *tertiary* formation of lime and clay and sand, on which are the diluvial deposits, and we come to the comparatively recent epoch when man had first his creation and abode upon the earth.

So would the rolling molten mass have formed its surface, when thrown off as a planet from the revolving sphere that had previously discharged the planets which are beyond the earth. The specific gravity of the matter must have determined the lighter to the higher positions, and as these successively cooled and hardened, we should have the mica schist, the gneiss and granite in their places, and the crust then of sufficient thickness, and the temperature of sufficient coolness, that water and an atmosphere might there be formed, and the life-force be introduced by the Creator, and the germs of plants and embryos of animal being be given to material nature. Disintegrations and decompositions and subterranean convulsions must perpetually occur, affording materials for new deposits, and which may have many alternations of submersion and upheaval, and thus ranging the strata and their varied fossil remains as the geologist now finds them. These subterranean forces must have often ruptured the whole crust, and turned up the edges of the strata, and formed the mountains and valleys, and exposed the granite, gneiss and mica as they now appear. Oftentimes the molten matter beneath will have been forced through the fissures of the granite and overlying strata, and cooled into the forms of trap and basalt as now found in their localities.

So, manifestly, with the revolving earth. But the Moon, with no rotation on its axis, must cool and harden with no wrapping of its layers about it by its own motion, and its surface must consequently be, as telescopic observation finds it, broken into sharp hills and mountains, and these mountain crests the volcanic outlets and craters of the escaping fires. The body of the moon must thus long since have lost much of its inner heat, and its gases become condensed and fluids absolved mainly into its own substance. Water and an atmosphere, if any, must be too low for ordinary telescopic observation.

Observation also determines the fact of an atmosphere in the twilight of Mercury and Venus, and not only an atmosphere, but the vapor of clouds and polar snows upon Mars, and a very dense but little elevated atmosphere above the surface of Jupiter; and as these with all the other planets rotate on their axes, so doubtless the same laws for binding down the internal heat beneath the overwrapping strata, that is found in the earth, prevail in all the other planets; while the fact that no satellite seems to have any rotation, but only to turn its side once to the primary in its revolution about it, as being in its revolution once above and once beneath it, would lead also to the conclusion that for these no atmosphere nor water need be expected to reveal themselves to our observation. The moons of all the other planets must be older than ours, and their cooling and absorption of vapors must have left an atmosphere on them, even less elevated than on our satellite.

Thus the great physical facts, through all the varied fields of natural observation and experience, are found to be completely bound in laws that are necessarily determined for them in the eternal principles which condition them.

13. Laws of Stellar Distribution.—On any clear night, there is observable a very perspicuous zone or belt of a curdled and in some places partially interrupted silvery white light, which as a broad bow spans the heavens through our whole hemisphere, and which is thus known as the *galaxy* or milky-way. The same belt continued spans the southern heavens, and is thus a great circle through the whole heavenly sphere. It makes an angle of

about 40° with the ecliptic, and its plane cuts our globe at an inclination of about 63° to the equator, crossing the equator on each side of the globe at about 10° east of the equinoctial points. It does not quite equally divide the heavens, but makes the two hemispheres proportioned as 8 to 9, with the vernal solstice in the smaller area. This galactic circle may be conceived as having its poles at the opposite extremities of an axis passing through the centre perpendicular to its plane. The northern galactic pole will then have its position in the constellation Coma Berenices, and the southern galactic pole between the tail of Cetus and Apparatus Sculptoris.

When this broad circle is viewed through a large telescope, the peculiar white appearance is seen to extend itself to some distance on each side, making the milky-way 6° or 7° broader in the heavens than when observed by the naked eye. Its narrowest and brightest portion is in the southern hemisphere near the constellation of the Cross and at the hind feet of Centaurus, being there about 3° in breadth. Its broadest undivided portions are in some places 15°, and where there is the broadest separation in the northern hemisphere, the whole width from outside to outside of the two paths is 22°. There are several breaks and interruptions and bifurcating separations in its course, but the most remarkable division begins in the southern heavens near Circinus and the fore feet of Centaurus, and one branch runs up and loses itself near the foot of Serpentarius, and the more southerly branch passes through Aquila, Sagitta, Vulpecula, irregularly but uninterruptedly to Cygnus. The lost northern branch also recovers itself in Aquila, and comes up to meet the southern branch in Cygnus. The bifurcations in Circinus and Cygnus are about 130° apart, and through the main distance both parts continue nearly parallel, though slightly converging from the mid-point.

The whole belt is very satisfactorily determined to be the shining of thickly clustered stars too minute to be distinguishable by the naked eye, the blended light of which makes this white circle across both hemispheres. A telescope of high magnifying power resolves very much of this belt into distinct stars of the smallest magnitude, and though very numerous yet may be carefully counted in the field of the telescope, while other parts, though distinct, are yet so finely powdered that like close grains of sand they cannot be numbered. Some parts of the milky-way appear to be filled with these minute stars, which though fully resolved do not leave any appreciable spaces separating them, and at other places the grains grow thinner and the dark unoccupied spaces open between them. There are other regions which to the highest glasses are still unresolvable, and though piercing to more than 2000 times the distance of the nearest stars, and from whence it would require that the light should be more than 12,000 years on its passage to us, yet is the depth to which stars beyond stars are here placed wholly fathomless. In some parts the stars of the first magnitudes appear to lie on a background of the smallest resolvable stars, or on a ground of unresolvable brightness at a great remove behind, and nothing within the broad space between; and in other parts the successive magnitudes seem to lie regularly stratum behind stratum, filling the whole depth to the most fathomless distances.

From each galactic pole up to the circle the spaces between have been carefully gauged in both hemispheres, and in various lines of the galactic meridians, that it might be determined how the stars are distributed relatively from these galactic poles to the middle of the circle. The whole breadth of 90° was divided into zones of 15° each, and the field of the telescope passed up through them in one successive galactic latitude after another, designing to cover at each remove a circle of 15′ diameter. About 2,300 careful gauges were made by Sir J. Herschel in the southern heavens, and a similar observation had been made by Sir W. Herschel of the northern heavens, and an extended analysis of these observations by Prof. Struve determined the following results of comparative stellar distribution, from the galactic poles upwards through these successive zones.

Northern Galactic Pole.		Southern Galactic Pole.	
90° to 75° =	4,32	90° to 75° =	6,05
75° to 60° =	5,42	75° to 60° =	6,62
60° to 45° =	8,21	60° to 45° =	9,08
45° to 30° =	13,16	45° to 30° =	13,49
30° to 15° =	24,09	30° to 15° =	26,29
15° to 00° =	53,43	15° to 00° =	59,06

At the galactic circle, 122,00.

The increase maintains an astonishing degree of regularity and uniformity in each hemisphere, the southern being invariably a little advance upon the northern, but constantly in very fair proportion from zone to zone respectively. The galactic poles have few, and the galactic circle has many, stars.

There are some very striking and important facts to be

noticed in this augmentation of stars through the higher galactic latitudes, viz. that the larger stars up to those of the 8th magnitude had no perceptible increase as the latitude was elevated; that the stars of the 9th and 10th magnitudes regularly increased from about 30° on each side of the galactic circle; that stars of the 11th magnitude began to increase soon after leaving the galactic poles; and that from the 12th magnitude and upwards the increase was striking and constant from the start. The stars at the galactic poles are thus in much greater proportion of larger to smaller than at the galactic circle. The larger keep their own number all the way, the smaller only make the increase. It is also a noticeable and important fact, that the stellar clusters and nebulæ are very infrequent, and almost none within the galactic circle, but in different directions these clusters and nebulæ are numerous at large distances from the circle. The most remarkable field of any in the heavens for these clusters and nebulæ, is in the neighborhood of and above the north galactic pole, where in about one-eighth of the heavens is found one-third of all the clusters and nebulæ.

The next in importance is the region about and above the southern galactic pole in the neighborhood of Pisces, so that there has been conjectured to be a belt of clusters and nebulæ at right angles to the galactic circle. But close and continued observation does not confirm the continuance of such belt of nebulæ transverse the milky-way, and only determines their existence to the fields above the galactic polar regions, the most numerous and extensive being found at the northern galactic pole. Indeed that part of the region between the galactic poles that lies about Aries, Taurus, and the head of Orion, is almost wholly barren of all nebulæ. The catalogues of nebulæ give much the largest number to the northern portion of the heavens, making 2299 for the northern, and 1239 for the southern hemisphere. The stellar clusters reverse this order, being 152 north and 230 south.

The most remarkable patches of clustered nebulæ are the larger and smaller Magellanic clouds, or *Nubecula major* and *Nubecula minor*, which circle about the southern pole at from 26° to 30° from it, and about 12° distant from each other. The first covers a space in the heavens of 42 square degrees, and the latter 10 square degrees. The base of these nebulæ is a wholly unresolved field of brightness on which are projected stellar clusters and single stars of varying magnitudes. Taking the entire Nubecula, there are in the major 291 nebulæ, 46 clusters, and 582 single stars; and in the minor there are 37 nebulæ, 7 clusters, and 200 single stars. The stars and clusters are doubtless much nearer to the point of observation, and stand far within the distance of the unresolved nebulæ that are in the background. It may be said generally of the stars in the southern hemisphere, that though fewer than in the northern, they are resplendent with a very appreciably augmented brightness.

Now this somewhat extended enumeration of facts, in reference to stellar distribution, can go but little way as data for determining, by any deductions therefrom, the general structure of the starry heavens. We look through into the dark open space, in some places of the surrounding firmament; again, we have stellar clusters of less and greater spacial area in which, though we may distinguish the exceedingly minute stars they contain, the resolved stars are so thickly studded together, that the spaces between are barely recognized; and then there are other considerable regions of fathomless depth, through which no assisting powerful instruments help the sight to pierce and determine that there is other than a perpetual heaping of stars one upon another.

How then the external limits of the starry universe are to be drawn, and what internal parts may be filled or empty, and what the relative shapes and bearings from each other any divisions of the whole may give, the ascertained facts are as yet far too scanty and partial to determine when left to the teaching of the facts alone. A pretty common opinion has apparently been admitted, that our system is plunged deep in the separate starry stratum of the milky-way, and that other distant nebulæ are strata as distinct and probably as large as our galactic stratum, and each forming as perfect a circle of silvery light as ours, to those who inhabit the worlds that are as deeply imbedded within it as we are in ours. Sir William Herschel early expressed such a conjecture, and this has been silently acquiesced in rather than adopted from any careful inquiry, though it is manifest that this early conjecture of Herschel has lost all real support from its author by his subsequently disproving the supposed facts that had given rise to it. The field of the milky-way has never yet been other than very partially fathomed; large portions are wholly unresolved by the most enlarged modern reflectors. The outskirts of the milky-way may be as remote from us as any unresolved nebulæ, and we cannot from any observed facts yet conclude that it is a separate and independent stratum.

But when we have attained the principle that must determine the structure of the stellar region, and necessarily fix its locality within definite limits in the universal sphere, and thereby apprehend what must be the aggregate localities of the starry worlds, we may then bring all the facts we have immediately under the circumscription of such determining principles, and therein find the accordance of facts and principle much sooner, than from any induction of facts alone we could hope for the suggestion of an hypothesis that would bear the test of general observation and experiment. The combination of forces, that must result from the universal hemispherical pressure and the central impulse of the perpetual generation of the conjoined antagonist and diremptive forces, necessarily excludes all world-formations within the limits of two spheroids formed on the semi-diameters of the universal sphere, and determines their construction in successive layers or strata in the circumference of such spheroids, and thence externally in the diminishing arcs of increasing circles, till the compound forces find their equilibrium out near the surface of the universal sphere.

We will, then, again take the general diagram that represents the stellar structure of the universe, and the few facts already attained will enable us to so locate our solar system, and our terrestrial stand-point in it, that we may make these facts altogether to conform to the celestial appearances of the most particular and the most extended astronomical observations.

We have the bisection of the universal sphere through

its polar diameter in the circumference M F W D, and the two spheroids C H D and C I F about the semi-diameters C D and C F. The stellar region therefore occupies the space exterior to the spheroids C H D and C I F, and ex-

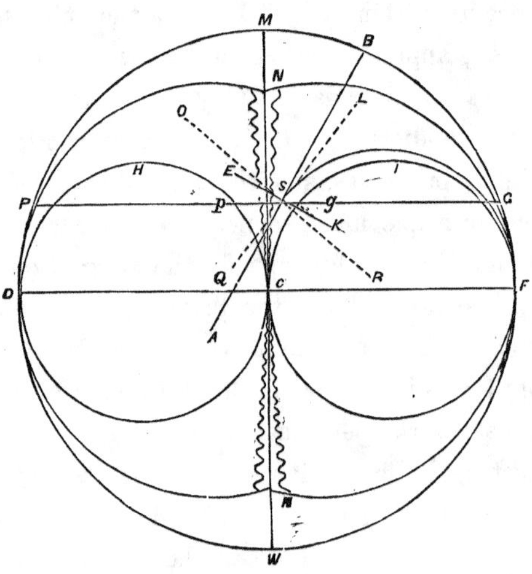

tends out indefinitely toward the circumference of the universal sphere till it finds its equilibrium of hemispherical pressure and central impulse, say, in the general spherical strata N F and N D. The equatorial plane M W will then indicate the apparent place, in its extremities, of the great galactic circle. Now this galactic circle has the earth's equator inclined towards it about 63°, and which must make the plane of the earth's equator, say, in the line E K, and consequently the polar axis in a line perpendicular thereto, viz., B A, with B its boreal and A its austral point. But the galactic circle divides the apparent heavens somewhat unequally, in the proportion of about 8 in the southern side to 9 in the northern, and which must place the stand-point of observation in our system so much to the right hand of the universal equatorial plane, or plane of the galactic circle, say at S. The ecliptic must thus be about 23° inclined from the earth's equator, and about 40° inclined toward the galactic circle; or, if we take the mean between the earth's orbit and the plane of the sun's equator, as the general orbital plane of the whole solar system, we shall have about 27° of inclination from the plane of the earth's equator and about 37° of inclination towards the galactic circle, making this general orbital plane of the solar system to be in the dotted line O R. This orbital plane of the system must, again, have been at right angles to the tangent of the spherical stratum out of which the solar system was constituted, and about the line of which the central impulse must have made it to revolve, and which will make that tangent to be the dotted line L Q, and the stratum to be the arc of the sphere S F. And since this also exactly conforms to the demand for a diminished arc of an increasing sphere, proportioned to the distance from the centre C relatively to the circumference of the sphere C I F, we have all the perfect coincidences of fact and principle through the whole operation that can belong to nothing but true science.

The point S may be any one in a circle of points at the same distance from the centre and also from the galactic plane on that side, and all such points will have the same general relation to the universal stellar structure, and from each as a stand-point the heavens will have the same general appearance. On the opposite side of the galactic plane there must also be another circle of points in the same relative position to the stellar structure from that hemisphere, and from which the heavens must have the same general appearance, except as the galactic circle must stand in the opposite relation, and the orbital movement of the systems must be in the opposite directions. But many further facts perfectly coincide with the principle which determines this position for our solar system. Beside the unequal apparent division of the heavens by the galactic circle, from a point about 5° removed from its plane where our system stands on the right hand side, there must also from this position be the apparent bifurcation and division of the milky-way just as our heavens present their phenomena. The waving lines diverging from the centre on each side of the universal equatorial plane, as given in the figure, represent in their opening the vacancy that must occur in the stellar distribution quite around the universal sphere, on account of the gradually diminishing forces in operation as the distance from the centre increases, and such opening must appear as the galactic separations are seen to be from our point of observation. Each diurnal revolution must bring the observer under such opening, and if in the northern hemisphere, then the northern place of bifurcation must appear above the horizon, and if the observer be south, then the southern place of bifurcation must appear, and if on or near the equator, then both the northern and southern places of separation will be seen with all the space between them. At some part of every diurnal revolution must also the observer lose sight of the galactic separation, for he must daily pass under that portion of the galactic circle whose direction will be within toward the universal centre and where the separations must be too slight to appear.

The hemispherical pressure and the central outworkings in composition must determine the thickest material to be generated and located near the regions on each side of the universal equatorial plane, and thus each stratum will there have the stars the nearest together and becoming more sparse backward toward the universal poles; yet such frequency of stars in their respective strata will not be the chief occasion of the white galactic light, but the much deeper space and thus the many more stars that come in the galactic plane from our stand-point on the earth, and the range that the earth's position must give to them in their general distribution to secure such concentration in the line of vision. If we take a field of vision from the earth's position with the distance S C for a spherical radius, it is manifest that the direction in which the stars will appear to be constantly condensing from all sides will be in the plane of the galactic circle, and the same will also be true if we increase our spherical radius to any telescopic distance, S A or S B; the position of the earth is such in the great stellar structure, that the plane of the galactic circle must be that to which the lines of vision are on all sides converging, that they may pass through the thickest portions or the deepest gauges of the stellar region. The narrowest portion must be in the direction S C, and which will bring the line in the region of Charles' Oak near to the Southern Cross, and the broadest portion must be in the direction S N, which will give the middle of the galactic separation, or the region of Aquila; and to such a result the facts themselves conform. The brightest part of the heavens will be that in which the proportions of number and magnitude of the stars combine to give the greatest light, and which will not necessarily be in the thickest part of the milky-way. Some of the brightest portions of the heavens are lighted up by stars of the first magnitude occurring near together and also in near neighborhood to the milky-way, as in the region of the Southern Cross, the portion from Orion through Canis Major and Argo, and also the Altar and the tail of the Scorpion, and in the northern hemisphere the parts about the bifurcation in Cygnus, but these bright stars of the first magnitudes are doubtless to be apprehended as comparatively near to the earth, and projected by the line of vision upon the circle of the galaxy beyond them, or the spaces adjoining. The least illuminated portion is that part of the milky-way between Monoceros and Perseus, and which will stand opposite to the galactic separation, or towards the interior of the universal sphere.

But the most striking conformity of facts and principle, given in this position for our system, is the relative proportion of the stars in numbers as the observation proceeds from the galactic poles on each side up to the galactic circle. We have above seen that the gauges from the northern galactic pole to the circle, are invariably somewhat less than those from the southern galactic pole to the circle, though the ratios of increment are very similar in each, and the position of the system must make such appearance necessary. A gauge at the distance of a spherical radius S g in the direction of the northern galactic pole, up to the galactic circle, must perpetually pass into an increased thickness of the stellar strata, and thus a continually augmenting number of stars must come within the vision until the milky-way is reached in the meridian.

And so also with the spherical radius from S p, in the direction of the southern galactic pole, a similar augmentation of thickness in the stellar strata as observed from our position must occur, as the gauges approach the milky-way in the meridian from the southern galactic pole. But inasmuch as the earth's position in its system is a little on the northern side of the galactic circle, the thickness of the stellar strata and consequently the number of the stars in the vision must always be proportionally greater through the lines of the southern gauges.

And still further, this increase does not occur in stars larger than the 8th magnitude, thus manifesting that the stars at the distances of S p and S g must be of the 7th and 8th magnitudes. Within the spheres formed by the radii S p and S g, the stars must increase in magnitude to the 1st as nearest to S, and as those of the same magnitude will on an average stand out on all sides from S in the same thickness of stellar strata, so the gauges must make on all sides the same numbers for the same magnitudes. But it will be observed in the figure that an increase of radius beyond S p and S g to reach the stars of the 9th and 10th magnitudes, the gauges must meet the arcs of the lower stellar strata C H D and C I F quite above the line of the galactic axis, and within about 30° of the galactic circle, and from thence the gauges can make some but only a slight increment up to the meridian. And then again, extending the spherical radius to reach stars of the 11th magnitude, the gauges must meet the spherical arc still further from the galactic circle, and their ratio of increment must be proportionally conspicuous; and from thence to the smallest telescopic magnitudes of the 17th and 18th, we come to the spherical radius at the distances S G and S P, where the gauges must meet the spherical arcs on their opposite sides, and the stars of these smallest magnitudes are found even at the galactic poles, and their increment is perpetual from the pole quite up to the galactic circle.

It is also in this manifest, how the stars of larger magnitude may be found with large vacancies between them and those of much smaller magnitudes. By looking through the telescope in the direction of either galactic pole, the stars in its field will be visible from S g and S p, as stars from the 1st to the 8th magnitude, but the space will then be vacant till we reach the stars on the further portion of the spherical arcs, and of the smallest magnitudes about the galactic poles. As the telescope shall be elevated above the galactic poles toward the meridian, the arcs will be struck at higher points and the vacant spaces will grow perpetually smaller, until they vanish quite away in the culminating point, and the stellar strata will be from thence in solid continuity.

The stellar clusters and nebulæ also quite conform to the conditions demanded in this relative position of the solar system. If the nebulæ are but stellar clusters unresolvable by telescopic power because of their greater distances, then it must follow that the unresolved nebulæ must be in those portions of the stellar structure that admit of great distance. The region most fertile in stellar clusters and nebulæ is that about and above the north galactic pole, in the constellations Leo Major, Canes Venatici, Coma Berenices, and the head and wings of Virgo; and in the region about the south galactic pole with a northern direction from it, in the constellation Pisces, there are also frequent stellar clusters. By large magnifying glasses these are mostly resolvable, and therefore must lie within distances no greater than those of the galactic poles. The nebulæ in the sword of Orion and another in Andromeda's girdle, are barely if at all completely resolvable, and they lie, the former pretty nearly over the earth's position, when the southern side of the galactic circle is on the meridian, and may therefore be at the distance of a spherical radius S B, and the latter nearly in the direction S O, and may therefore be at any distance admitted by the whole thickness of the stellar strata in that direction.

The very remarkable nebulæ, or clusters of nebulæ, in the extreme southern hemisphere, called Magellanie clouds or Nubecula major and Nubecula minor, have been above described. The stars and stellar clusters, which appear in the line of vision projected upon the nebulæ as their background, must have an intervening position, and the nebulæ themselves, which are hitherto wholly irresolvable, must be among the objects furthest removed within that sphere that the light penetrates to reach our earth. Their positions are near the south terrestrial pole, the Nubecula major nearly in its direction, and the Nubecula minor on the opposite side of the pole from the milky-way, and an inspection of the diagram at once convinces that the stars and stellar clusters have their positions within the distance S C, in the direction of the earth's axis, and that the nebulæ themselves are stellar clusters on the inner stratum of the spherical arc below the south pole, as here presented beyond the distance S A.

These many conformities of facts and principle, confirm

the position of the solar system and the law for stellar distribution as we have given them.

14. Laws of Life.—The most general laws only need here be noticed, for the general principles of the life-force have not been sufficiently determinative of the facts to make any minute conformities extensively observable. The life-force is a spiritual activity and retains its simplicity in its incorporation with matter, and can, therefore, never itself become phenomenal. Material forces only appear, and the life-force becomes known only as it registers itself in the material forces which it assimilates. Our insight of it is only in the conception that it meets the simple activities of the material forces, and thus becomes itself a force by dissolving and interpenetrating other forces and using their component simple energies. When thus using and assimilating the elemental activities of other forces, the matter incorporated by it is quick and organic, but when the assimilating activity is withdrawn, the material forces run on again their own conditioned changes.

The spiritual activity in the life-force has within its first incorporation as a germ, the potential forms of its full maturity, and each germ uses material nature in its own way for building up its own forms in the matter it incorporates. This inherent formative power adapts itself to its wants, and sends down a tap-root in the vegetable, or heals a wound and mends a broken bone in the animal, when circumstances demand it.

The vegetable life can use for its body the unorganized material forces, and such rude matter as has never come into any organization is here first made the tabernacle for the living spirit. This spirit, as a mere *vegeta* solely, incorporates without any sentient activity, and uses certain appropriate unorganized forces as the elements which it assimilates and takes into its own organism. And yet such matter as has already been used in vegetation, the old cast off remnant and dissolved portion of decayed vegetation, seems the most readily to be made the nourishment of other plants. The animal body is wholly fed by that which has once been vitalized, and must live and grow by incorporating vegetable matter, or the bodies of other animals, and never feeds on unorganized substances.

The law of sex, which the idea makes necessary to the propagation of the species, binds up the facts in vegetable as well as in animal generation. The fructification in plants is through the sexual flowers, and these may be found sometimes of both sexes on the same plant—the flower and fruit on different places in the stock, or sometimes both flower and fruit set in the same calyx,—or the flowers are invariably found, in other varieties, with opposite sexes on different stocks. Again, among animals, the propagation is sometimes through an impregnation of an ovum, and which by incubation produces the offspring in an oviparous birth, or the pregnancy takes place in a teeming womb and the embryo is produced in a viviparous generation. The law of sex is universal in all organized material being, and the great difference between generation and growth is, that the growing development is an assimilation of matter in the germ which passes back and states itself in the stock, or is taken up and incarnated through the circulation; while in generation there is a new organism begotten, that ultimately separates itself from the parent stock, and passes on in its own growth by an independent development.

All different varieties of the same species may procreate, but the general law is that different species shall not mingle in procreation. The facts sometimes give an offspring from nearly allied distinct species in the same family, but such hybrid generations either never propagate, or only through short successions by a recurrence to one of the original species. Nature soon excludes a blended progeny.

We finish this chapter, and thus conclude our whole work, by a short statement of the LAW OF PHYSICAL ENERGIES, which the principles of the forces in nature determine, and enable us by a rational insight very completely to apprehend as necessarily inclusive of all static and dynamic agencies.

The sources of mechanical power, of which in experience any one can avail himself, are sufficiently definite to be exactly and comprehensively classified. All mechanism may be traced to one of the following sources of moving energy, viz., 1. Weights. 2. Flowing currents. 3. Heat and cold. 4. Combustibles. 5. Magnetics. 6. Electrics. 7. Muscular action.

By a closer careful observation in experiment, it becomes manifest that a number of these are convertible with others or may be included within others. The impulses of flowing currents, whether of fluids or gases, may be combined in some cases with weights, and all be classed as gravity, or in other cases with heat and cold, and thus all be classed as expansibility. Running streams and flowing tides and atmospheric pressure come under the former; steam, congelation and crystallization come within the latter. From what we have seen also of the combination, in congelation and crystallization, of the matter congealed or crystallized with the latent heat of fusion, or with the heat set free in chemical dissolution, we may exclude cold as any positive agent and ascribe all expansion to heat.

Again, the force of combustion may also be referred to the one source of heat. Experiments prove that by the action of solar light on the green parts of plants there is effected a deoxydation of carbon and hydrogen from carbonic acid and water, and that thus what is combustible in all vegetation is the product of radiated heat in combination with the material forces. This combination constitutes the fibrous or woody substance in plants that is known as combustible. The incorporated solar heat lies latent in the substance, and the fire of burning wood, charcoal, or mineral coal, is only solar heat reproduced in the liberation given to it by the dissolving combustion.

We may thus reduce our physical agents to, 1. Weight. 2. Heat. 3. Magnetism. 4. Electricity. 5. Muscular action.

But, as we have found the comprehensive antagonist-force to be both gravity, magnetism, and electricity, and therefore the whole to be but one agent in different relative positions, we have then, in reality, only the three sources of physical energy, viz. Gravity, Heat, and Muscular Action. But muscular energy may now be very safely taken as having its whole source in the heat-force. Careful experiment has well established the fact that the animal heat, the secretion and excretion of the fleshy tissues, and the muscular energy, find their respective equivalents in the oxydation and assimilation of the food used. There is a portion of heat given off in the organism and radiating out in the atmosphere, a portion combined by vital chemistry in the incarnating or flesh-making process, and another portion exhausted in muscular action. The whole process of the transformation of the constituent heat in the food into animal heat, fleshy fibre, and muscular energy, cannot as yet be brought within experiment; nor can the due proportions of the heat-force in the food be precisely determined for these three results, in order that they may be most perfect on the whole in the animal economy; but that an increase in either one, will find its equivalent diminution in one or both of the others, may be taken as sufficiently established. When digestion and assimilation goes on in a state of general muscular rest, the oxydation of the blood in the lungs, and the determination of the circulation in the arterial currents, secures a perpetual combustion, and producing of carbonic acid as the result of the oxydation, and thus a portion of heat is communicated to the blood and imparted to the whole organism. When strong muscular exertion occurs, more heat is necessarily demanded and used in the muscular contractions; and while the arterial circulation is quickened, the heat of the animal system is proportionally elevated and worked off in the muscular labor, and a proportionally larger new supply of food must be procured.

From pretty careful estimates it has been concluded that about $\frac{1}{5}$ of the heat liberated from the food in oxydation may be regularly returned, by a horse at labor, in weights raised or resistance overcome. It cannot be said that all the food is at any time completely oxydated, but about $\frac{4}{5}$ of what is chemically supplied in the whole animal laboratory must be used in the other agencies of the animal economy, and if any more than about $\frac{1}{5}$ be exhausted in muscular action, the animal organism will be injured. It is doubtless fair to infer that where the muscular exertion is too little called regularly forth, the redundancy of the heat liberated and entering into the assimilated fleshy fibre, or into the chemical process of dissolving the worn-out living tissues that the excrementitious portions may be thrown off, must induce the deterioration of the animal system that always follows the neglect of proper muscular exercise. No economical expedients can secure more than a certain amount of work done for a certain amount of food consumed, and no sanatory regulations can secure health, but by the supply of so much food, and so much of its liberated heat used up in muscular activity.

It will thus follow, that the vital agency is no source of physical, mechanical energy. It is not the source of any new power, but in its interaction and incorporation with other forces, as antagonist or diremptive, it uses such forces and subserves its own wants by them. The heat-force, especially in the matter that the vital activity interpenetrates, is made by the vitality to work for it. As a sentient life the activity may excite the muscular irritability, but the muscle has no mechanical force to be applied, except as it is itself supplied by the dissolved food in the heat-force thence imparted. The formative energy inherent in the germ has secured the fibrous arrangements and attachments of the muscle, and the sentient irritation that uses the conveyed heat-force secures, by the muscular contraction, the mechanical effects determined in the prearranged attachments of such muscular pulleys to their strong bony levers. We have therefore, in this exclusion of the vital agency from any direct mechanical energy, only the forces of Gravity and Heat as comprising within themselves all the physical powers in nature.

This comprehension of all mechanical force and movement within the two original and constitutive forces of nature, antagonist and diremptive, is a striking confirmation that our Cosmology has been made to rest on the basis of a true and valid science. The principle of the generation of the material universe involved the agency of these two forces, and needed none other. Spiritual Activity finds occasion to go out in its energy in the origination of both counteraction and diremption, and in the combination of these, all that is space-filling and time-determining has its existence. And now, when we come to examine the actual forces operating anywhere within the created universe, and make an analysis of all the mechanical powers we can extract from nature, we find them all resolvable into weight and expansion, or gravity and heat, which are only the pressure of antagonist activities and the disparting of diremptive activities. Nature needed nothing more for its own existence; nature uses nothing more for its onward development; nature yields nothing more to human solicitation or extortion.

And the amount of this mechanical energy of both kinds is wholly incapable of either augmentation or annihilation, but by an absolute and supernatural agent. The only way of coming in and going out is through the great central Spiritual Activity. New creations flow only from that source; annihilations of old existences can only go up back again into that source. Nature's gravity and nature's heat have perpetually the same measure in the aggregate, and both equilibrate each other; and these intelligence may use according to their necessary laws, but all derangement of the necessary laws of nature would itself subvert all intelligence.

APPENDIX.

RATIONAL COSMOLOGY ACCORDS WITH THE MOSAIC HISTORY OF CREATION.

In the first two chapters of the Book of Genesis, is a general history of the creation, and though very concise, yet is the record very methodical and comprehensive. The first chapter and the first three verses of the second give the process in detail through six successive days, and the cessation or resting from the creative work on the seventh. The remainder of the second chapter has a more desultory account of some promiscuous items of the work together with some of the circumstantial dealings of the Creator with man in his primitive state. The creative work in general, and the making of plants and herbs antecedently to any rain upon the earth, together with the formation of the first man from the dust of the ground are mentioned in verses 4–7; the planting of a garden and putting the man into it to dress and keep it, and the prohibition to eat of the tree of the knowledge of good and evil are narrated in verses 8–17; and then the making of woman from a rib of Adam, and giving her to him in marriage, are described in the last part of the chapter, verses 18–25.

This inspired record is to be understood as God's representation of his own work, and containing the truth so far as the history goes, whether it be supposed to have been originally composed by Moses under inspiration, or by divine guidance adopted by him from some earlier composition, and whether, again, it be assumed that Moses was possessed of the full meaning of the communication or not. This record is not given in scientific form nor with philosophical precision and method, but, as Dr. Lewis has ably and satisfactorily shown in his Six Days of Creation, the facts are presented as they would appear through the medium of the senses. Principles and laws as given in thought are not noticed, and only phenomenal representations are made as they would appear alike to the unreflecting and the scientific scholar. The language is designed to carry a common meaning to all readers in every age.

Natural science, especially in the fields of Astronomy and Geology, has attained conclusions which have seemed in some cases to be in conflict with this Bible record. Philological interpretation has been modified in various ways to meet these difficulties from science, and by looking at the Scripture account as intended to give a picture of facts for the sense, and interpreting some words by usage in other places of the Scriptures with a less common meaning, the discrepancies have been much relieved, and science and the Bible surprisingly harmonized and made to be corroborative of each other. A correct Bible philology and a true natural philosophy must doubtless give facts in unison, and where their facts seem in any measure as yet to be contradictory, a more complete investigation will at length secure a thorough communion.

But there are not only the difficulties introduced from the discoveries of science, and which more complete investigations must remove, there are also some inherent difficulties in the Mosaic record itself, and which are induced by apparently irreconcilable incongruities in its own statements, and which no induction of facts from human experience seem able to reach and relieve. The double creation of light, or the making of light antecedently to the making of the sun, the creation of plants and herbs before the sun was made, the mist that covered the earth before there was any rain, etc., are of this description and still leave their perplexities in spite of labored attempts at explanation. And now, while the principles and laws attained in the foregoing cosmological investigation will be found fully to harmonize their determinations with the phenomenal representations of the Bible and the general deductions of natural science, and even more clearly to expound and harmonize their separate facts, there will moreover be found this incomparable advantage from the cosmology, that it will remove these apparent incongruities in the statements of the record, and show the facts to be as the statements from the necessity of the case. We shall need only to follow the creative process step by step through the successive days in the Mosaic record with the cosmological principles attained constantly in mind, and the accordance of each, and the explanation of one by the other will continually appear.

The philological question whether the Hebrew word for "created" in the first verse means creation in the sense of origination, or only a new fashioning of old materials may be decided either way without prejudice to the main facts in the history. If the first view be taken, then "the beginning" is the first origin of material existence, and if the second be adopted, then "the beginning" is subsequent to the origin of matter, and is the commencement of that work which remodeled our world out of old materials, and placed it in sensible communion with other worlds around it. The last view may be consistent with the creed of a divine origination of all matter, though it exclude the inspired communication of such a fact from the Mosaic account. After the first verse, both views would proceed in the same way. But the cosmological principles attained enable us to go back to the very morning of creation in the origination of matter itself, and with this light we can hardly fail to recognize a divine intention in the Mosaic history to give to us an account of the very beginning of material existence. The first verse of the Bible may well be applied to what we recognize as the cosmological fact, that God in his pure spiritual being put forth his simple activity in an antagonist action originating force, and thereby took and filled and held a position in space, and this filling of a place with a force that must exclude all other forces from the same position was the origination of substantial matter, and thereby the occupying of so much space with what was thus made by him to *stand out* from him, or to *exist* distinct from his being, though from the first and ever dependent on his agency. The work of creation, commencing in this first point of counter-agency, must pass on through all the process of generating the universal sphere, and bringing it into a fluid state by the permeation of the heat-force that just held loose every molecule in the primitive ether, and then sending the continual stream of the combined central forces in composition with the counteracting hemispherical pressure, as we have above carefully traced, and thereby thickening this primitive ether to a chaotic state of chemical forces that became a resisting material pushed and driven into myriads of separate wheeling spheres, and those spheres sending off from them in their revolutions, each its own planets and their satellites. The perpetuation of the central counter-agency must, at length, necessarily result in this making of countless distinct systems and their orderly arrangement in the universal ethereal sphere.

In all this process the passing of a time could only be determined by the movement that in succession was filling out this universal sphere, and pushing its divided portions into wheeling spheres, and forming these spheres into separate systems. No cyclical measure can be applied to estimate the duration, and give a definite period to the process. But up to the point when, in our solar sphere, our planet Earth was thrown out by the projectile tangential force, and sent in its revolution and rotation to condense and round itself into a solid globe, may we apply the declaration and include all the work and the time denoted in this first verse, "In the beginning God created the heavens and the earth." The whole material for the heavens and the earth was here brought into being, and other systems together with our solar system were in process of formation, and thus the creative work had its beginning.

And here, because the process in one world would be substantially the process in all worlds, and especially because it was designed for the intelligent inhabitants of our planet, the inspired history takes our earth as the stand-point, and describes the subsequent phenomenal facts and changes as if observed solely from this terrestrial position. The short general declaration in the first verse includes all that took place till the mass of our planet was separated from all else, and then that becomes the point of observation for all that follows. The next announcement in the second verse tells what was then its phenomenal condition. Just as the molten fragment had been thrown from the periphery of its wheeling sphere, itself shapeless without and as yet chemically uncompounded within, and flying off into the abyss before light had dawned, no words can be more forcefully graphic than these which inspiration has given. "The earth was without form, and void, and darkness was upon the face of the deep." But the forces which God had sent off in it, and with it, were working all through and about it. The revolving force threw the upper part of the mass over and forward, the last adhesions before they parted and the constant attractions of the parent sphere after the parting held the lower portions back, and this power, as if of the Spirit of God, brooded over the whole fluid matter, and brought the whole into form, and condensed its matter into chemical consistency. The earth became a globe of consistent compact matter, rotating on its own axle, and revolving in its determined orbit.

Thus finding the stable point of observation, the history begins an orderly and methodical process of narration. It divides itself into distinct notices of the successive events that come out from under the creative hand, and puts these into six successive days, when all is completed and the Creator ceases or rests from his work on the seventh day. These consecutive items follow each other just as the cosmological principles determine that they must, and just also as the geological discoveries affirm that they did, and the Scripture record is therein completely harmonious with cosmological truth and geological fact.

First Day. LIGHT. The earth was moving on in darkness, for though itself a fiery vaporous mass, and thoroughly permeated by the heat-force, still was this as yet so combined with the antagonist-forces that no radiations could go out in sufficiently modulated vibrations to become luminous. The primal sphere must diminish itself in the ejection of its planets and the condensing of its own matter, till it shall become a central orb of such dimensions that the ensphered ether about it shall gravitate towards it, and rest upon it, with sufficient intensity to give the concentric molecules consistency together as fixed spherical layers, and then the diremptive or heat-force must necessarily work itself out under this pressure, first in the polar and then in the equatorial directions from the centre and, in the consequent alternating prolate and oblate movement, the vibration we have before so thoroughly examined must begin and continue perpetual. When this ethereal gravitation has become sufficiently intense, and the surface of the central orb sufficiently contracted to secure the requisite breadth and rapidity to the vibrations, then must first the phenomenal light be given. This was the point in the creative work when God said, "Let light be. and light was." The period for this must have been long after the discession of the Earth as a planet, and quite probably even Venus was born in darkness, but ere Mercury became separated from this central body the luminous vibrations had doubtless commenced, and the morning-twilight of our solar system dawned on its hitherto cheerless worlds.

The first appearance of light would be faint, and grow brighter only as the condensing central orb grew less, and the pressure of the ethereal gravity upon its surface greater, nor for a long time could it be other than a pale effusion insufficient to give to an observer upon the earth the defined outlines of the orb from whence it came, nor by reflection the outlines of the revolving planets. As the earth turned on its axle, there would at length come the degree of light that should permit the observer to determine the direction from whence the radiations came, and thus to distinguish between the diurnal and nocturnal successions, and in this "God divided the light from the darkness. And God called the light day, and the darkness he called night." In this terminates the first historic epoch in the creative work, or the first creative day. That this could not have been a day determined by the terrestrial rotation to and from the luminous vibrations is manifest in that the creative day must have included countless numbers of the terrestrial natural day. The great wheeling sphere had been perpetually diminishing by its ejections and condensations from the space included in Neptune's orbit down to the space probably within Venus' orbit, and there the light was made by the ethereal pressure upon it and the radiating vibrations sent out from it, and this light augmented as the central orb contracted, and Mercury was ejected, but at the close of this epoch, and on until in the fourth creative day, the sun did not appear, nor was the light sufficient to give reflected vibrations that the moon should be visible. There were many terrestrial days and nights, without sun, moon or stars, in the first creative day of the Mosaic record.

Second Day. THE FIRMAMENT. At its first discession from the primal sphere, the mass from which the earth was composed was so intensely permeated with the heat-force as to be in a gaseous rather than a liquid state. Its volume then must have filled at least all the space within the moon's orbit, for its equatorial circumference must have been at that distance from its centre when the moon was thrown from it, and its daily rotation must have been in time the period of a lunar month. While, then, the great central orb was still further condensing itself after the earth's discession from it, and approaching that size which should induce luminous vibrations, and yet contracting more and more, the earth also was in the same way condensing and contracting its volume, and although during the first creative day no changes of a general appearance occurred in it only that light dawned upon it, yet in this second creative day it had cooled and condensed so far that an outer shell had formed over the consolidated molten matter within, and the strata wrapped over on the outside in the rotations had formed a stable crust, on which aqueous mist and vapor settled down and rested. As the coats of these cooling strata shut in more and more the radiations of the inner fire, the mist and vapors became more and more condensed and ultimately, in liquid form, the water accumulated at some depth over the even surface. The heavy mist in the thin atmosphere at first kept itself in contact with the earth, but the cooling crust and the accumulating water at length gave such density and buoyancy to the atmosphere, that the mist and vapors were lifted from the surface, and a separating space appeared between their lower stratum and the waters accumulated beneath. This interposed space of the cleared atmosphere grew gradually broader, and ultimately lifted the cloudy stratum so high that it appeared as a firm arch holding the mists above from all communion with the waters below. This is that of which the history says, "And God made the firmament, and divided the waters which were under the firmament from the waters which were above the firmament, and it was so. And God called the firmament Heaven."

This ancient conception was just that which the appearance presented. The phenomenal sky was a solid arch or dome standing firm in its place; above it, the waters in the clouds were treasured, and separated from the waters lying liquid on the face of the earth. The heavens are thus spoken of as being "spread out as a tent to dwell in;" and in the clear radiance of the day this firmament, or the sky, is said by Job to be "strong, and as a molten looking-glass." This space beneath the firmament would grow broader, and thus the heaven would grow higher as the atmosphere grew denser, but from the first it "divided the waters from the waters."

Third Day. DIVISION OF LAND AND WATER AND CREATION OF PLANTS. The superficial crust of the earth was still thickening, and the condensing vapors were then still augmenting the depth of waters upon it, and they must ultimately have covered the globe many miles deep. Inequalities of strength and thickness in the crust, and in the action of the superincumbent waters, and especially the contractions and condensations going on beneath must have made local elevations and depressions on the earth's surface, and occasional fractures must have occurred, and the mingling of the lower fires with the inflowing waters must have taken place, and many dislocations of the primitive strata must thus have been made, the most violent and extensive of which had upturned, and heaved out large fragments of the solid crust, and in this way hills must have risen and valleys sunk in the former level bed of the waters. The tops of continents like emerging islands then arose, and the dry land began to appear. This process went on until considerable portions had been elevated, and the waters had retired into the local depressions, and this work being accomplished, "God called the dry land earth, and the gathering together of the waters called he seas." The first uncovered portions must have been of limited area and of slight elevation, and the constant abruptions of the earth's crust must have made frequent alternations of uprisings and submersions.

Up to this point, the cosmological principles exhibit nothing but the play and product of physical forces, which in their combination constitute the material universe. The chaotic elements have come into chemical composition; metals, crystals, subcrystalline rocks, earth, water, air, and varied gases have been formed; but all is as yet inorganic, and the earth and waters azoic. The preparation is made, and the occasion given for the manifestation of organic existence, and here on the latter part of the third creative day we have in the record "And God said, let the earth bring forth grass, the herb yielding seed, and the fruit-tree yielding fruit after his kind, whose seed is in itself upon the earth, and it was so." The vegetable kingdom was thus introduced, and though at first marine plants and the succulent vegetation of marshes appeared, the work went on as the changing face of the earth prepared the way, and more mature species of shrubs and trees were made. The Mosaic history gives the introduction of the vegetable kingdom in its day, but new species were newly brought into existence all along down through the progress of the subsequent creative epochs. Geology finds fossil animals in the same strata with its earliest fossil vegetation, but the consideration that all animal life demands organic food, and that the plant was necessary as the absorbent of carbonic acid and the fixing of many gases in its growth, to clear the atmosphere for animal life, the evidence is sufficient that the plant must have preceded the introduction of animal existence. The earliest and frailest vegetable organisms may have failed to secure their fossil preservation, and only the more complete forms of a later creation have reached our age.

Fourth Day. HEAVENLY LUMINARIES. The earliest plants grew in the primitive mist and vapor, after the atmosphere had become sufficiently buoyant ordinarily to separate the vapors from the waters by the phenomenal firmament, but before it was sufficiently buoyant to sustain the dense rain-cloud; chap. ii. v. 5, 6. And also was their growth precedent to the direct warmth and light of the visible sun. The earth was too little cooled to permit other than a torrid climate upon all its surface, and the vibrations induced by the gravitating ether upon the central orb were sufficiently luminous for vegetable growth, before they became sufficiently intense to give to the sun a luminous envelope and make its face visible on the earth. But such luminous atmosphere about the sun must precede the introduction of animal life.

The comparatively recent ejection of Mercury, and the consequent condensation and contraction of the sun's volume, probably thereby the more suddenly occurring, brought its mass and the pressure of the gravitating ether upon it in such conformity, that an augmentation and accumulation of luminous vibrations upon its surface was effected, and henceforth light vibrated in radiations immediately from it, and not as before from the mere tension of the spherical layers occasioned by the gravitating ether towards it. The radiations thus bring the form of the sun with them, and represent it upon the organ of vision, "and the greater light that rules the day" was herein phenomenally made. This light reflected from the hitherto dark face of the moon caused it to appear, and "the lesser light that rules the night" was also phenomenally made. The planets that had with the earth been successively thrown off in the same system would shine by reflection, and the great orbs that were projecting their planets, and condensing their central matter into suns, in equal series with the process in our solar system, would emit their light as their direct vibrations should meet our earth, and thus phenomenally "he made the stars also." All would transpire within the same great epoch, and these luminaries would be for signs, and for determining cycles of time, and thus on the fourth day the lights in the firmament of heaven were made.

Fifth Day. FISH, FOWL, AND REPTILE. At this era the waters still abounded, and the tops of the continents only appeared, though the records of geology make it manifest that during the long time of its passing the waters had greatly receded, and the portions of dry land had become very much elevated and augmented. The third and fifth days' creative work must have been pretty closely consecutive, for the interposing fourth day's work was in reference to the lights of heaven, that had been in fact progressing and maturing from the first, and was phenomenally brought out between the third and fifth, but not necessarily separating the onward series of events from one day to the other. The order was, the creation of plants in the latest and strongest light that preceded the gathering of a luminous atmosphere about the sun, and that of fish and fowl with the earliest shining of this luminous solar envelope. The earliest fossils may thus well be of both plants and fish in the same rocky stratum.

Cosmical principles would also determine that the plant should not long precede the animal, inasmuch as the plant is for the animal and an immediate preparation for his introduction. And these principles also determine the order of the animal creation to be the marine and amphibian before the more mature mammalian family. Cosmological and geological teaching both conform to the Mosaic history. Nothing but a divine source could have secured to this early and unscientific record such a surprising harmony in so recondite a matter. The great heat of the earth was from within; the incipient forming of the sun's luminous atmosphere gave little warmth to the earth's surface. No thermal distinction of zones appear, and the fossils that would belong to torrid seas and islands are found within the Arctic circle. The species conformed to the temperature, and as the cooling process went on, whole species passed away and others succeeded; and of thousands that had their day in this fifth creative epoch, and have left their remnants in the rocks, none now exist among our present species. The molluscs, corals and fishes were followed by the amphibian races, and these by the reptiles and fowls. Sea and land often alternated over the same latitudes; successive extinctions of whole genera were made, and the long fifth day seems to have terminated in a catastrophe that abolished almost entirely its latest species.

The fruitful teeming of the waters with life, as geological fossils testify, is fully indicated in the Bible record. "And God said, Let the waters bring forth *abundantly* the moving creature that hath life, and fowl that may fly above the earth in the open firmament of heaven." The "great whales," of which our translation speaks, were the great Saurian monsters of this era, and of the race of reptiles rather than fishes.

Sixth Day. MAMMALS AND MAN. The broad continents were elevated and settled; the seas had attained to much their present boundaries; and then came the work of the sixth and last creative day. The more complete and perfect type of the animal organization had now occasion for its development. "And God said, Let the earth bring forth the living creature after his kind, cattle and creeping thing and beast of the earth after his kind, and it was so." The early portion of this sixth day was thus devoted to the making of the animals that bring forth a living offspring and nurse their own young. The expressions "creeping thing" and "every thing that creepeth upon the earth" refer not to the race of reptiles of the previous day's creation, but to the smaller mammalia, like the sloth and the mouse. The fossil records reveal that the Herbivorous preceded and the Carnivorous followed in the order of their production. In the palmy days of the brute race great numbers, and species of great magnitude, abounded. The Mammoth, and still more huge Mastodon and monster Magatherium, with ferocious lions, tigers, and hyenas, that exceeded any modern species, have left their fossil remains to testify how exuberant and robust was brute life in its most flourishing period. Cosmological principles and the facts found by science both determine this more perfect family of mammals, in the order of vertebrate animals, to this latest period of creation; and that the Mosaic record so exactly accords, coming down to us as it does from a period earlier than all philosophical teaching, is an abundant evidence that its testimony has the stamp and seal of one who was present when the things were done.

The closing of the Day that had introduced the most perfect forms of animal life ushered in the crowning product of creative power by the making of Man. In him came out, as a living organism, that which had all along been the archetype of every earlier animal form. A rational spirit was superinduced, and in the image of his Maker he had dominion over every living thing. The world found its end in him. In his production the last creative day closed, for the Creator's work was done.

Seventh Day. A SABBATH. All creative work ceased on the sixth day in the making of man. Nature henceforth goes on in the regular order of developed cause and effect, but no originations of new things take place, either of inorganic matter or of organic life. The worlds have reached their point of equilibration, and no occasions for new species of beings again occur. God, as Creator, ceases and rests from his work, and the world's Sabbath begins and lasts till the final conflagration. It has a hallowed relation to man's history; his probation, and coming retribution. God's dealings, all through the world's sabbatical epoch, apply directly to humanity; taking judicial recognition of his fall; introducing a promised way of redemption; applying a preparatory discipline of ritual education; bringing in the Gospel dispensation; and leading on the Church to the completion of Missionary effort in the Millennial reign of righteousness. The era of the world's Sabbath will be followed, immediately, by the opening of the endless day of rest in the Gospel Heaven.

FINIS.

Gems of British Art. 30 Engravings. 1 vol. 4to. morocco, ... 18 00
Gray's Elegy. Illustrated. 8vo. 1 50
Goldsmith's Deserted Village, ... 1 50
The Homes of American Authors. With Illustrations, cloth, ... 4 00
" " " cloth, gilt, 5 00
" " " mor. antqe. 7 00
The Holy Gospels. With 40 Designs by Overbeck. 1 vol. folio. Antique mor. . . . 20 00
The Land of Bondage. By J. M. Wainwright, D. D. Morocco, ... 6 00
The Queens of England. By Agnes Strickland. With 29 Portraits. Antique mor. . . . 10 00
The Ornaments of Memory. With 18 Illustrations. 4to. cloth, gilt, ... 6 00
" " Morocco, . . 10 00
Royal Gems from the Galleries of Europe. 40 Engravings, ... 25 00
The Republican Court; or, American Society in the Days of Washington. 21 Portraits. Antique mor. 12 00
The Vernon Gallery. 67 Engrav'gs. 4to. Ant. 25 00
The Women of the Bible. With 18 Engravings. Mor. antique, ... 10 00
Wilkie Gallery. Containing 60 Splendid Engravings. 4to. Antique mor. 25 00
A Winter Wreath of Summer Flowers. By S. G. Goodrich. Illustrated. Cloth, gilt, . 3 00

Juvenile Books.

A Poetry Book for Children, ... 75
Aunt Fanny's Christmas Stories. . . . 50
American Historical Tales, ... 75

UNCLE AMEREL'S STORY BOOKS.

The Little Gift Book. 18mo. cloth, . . 25
The Child's Story Book. Illust. 18mo. cloth, 25
Summer Holidays. 18mo. cloth, . . . 25
Winter Holidays. Illustrated. 18mo. cloth, . 25
George's Adventures in the Country. Illustrated. 18mo. cloth, ... 25
Christmas Stories. Illustrated. 18mo. cloth, . 25

Book of Trades, ... 50
Boys at Home. By the Author of Edgar Clifton, 75
Child's Cheerful Companion, ... 50
Child's Picture and Verse Book. 100 Engs. . 50

COUSIN ALICE'S WORKS.

All's Not Gold that Glitters, ... 75
Contentment Better than Wealth, ... 63
Nothing Venture, Nothing Have, ... 63
No such Word as Fail, ... 63
Patient Waiting No Loss, ... 63

Dashwood Priory. By the Author of Edgar Clifton, ... 75
Edgar Clifton; or Right and Wrong, . . . 75
Fireside Fairies. By Susan Pindar, . . . 63
Good in Every Thing. By Mrs. Barwell, . . 50
Leisure Moments Improved, ... 75
Life of Punchinello, ... 75

LIBRARY FOR MY YOUNG COUNTRYMEN.

Adventures of Capt. John Smith. By the Author of Uncle Philip, ... 38
Adventures of Daniel Boone. By do. . . 38
Dawnings of Genius. By Anne Pratt, . . 38
Life and Adventures of Henry Hudson. By the Author of Uncle Philip, ... 38
Life and Adventures of Hernan Cortez. By do. 38
Philip Randolph. A Tale of Virginia. By Mary Gertrude, ... 38
Rowan's History of the French Revolution. 2 vols. 75
Southey's Life of Oliver Cromwell, . . . 38

Louis' School-Days. By E. J. May, . . . 75
Louise; or, The Beauty of Integrity, . . . 25
Maryatt's Settlers in Canada, ... 62
" Masterman Ready, ... 63
" Scenes in Africa, ... 62
Midsummer Fays. By Susan Pindar, . . . 63

MISS McINTOSH'S WORKS.

Aunt Kitty's Tales, 12mo. 75
Blind Alice; A Tale for Good Children, . . 38
Ellen Leslie; or, The Reward of Self-Control, 38
Florence Arnott; or, Is She Generous? . . 38
Grace and Clara; or, Be Just as well as Generous, ... 38
Jessie Graham; or, Friends Dear, but Truth Dearer, ... 38
Emily Herbert; or, The Happy Home, . . . 37
Rose and Lillie Stanhope, ... 37

Mamma's Story Book, ... 75
Pebbles from the Sea-Shore, ... 37
Puss in Boots. Illustrated. By Otto Specter, . 25

PETER PARLEY'S WORKS.

Faggots for the Fireside, ... 1 13
Parley's Present for all Seasons, . . . 1 00
Wanderers by Sea and Land, ... 1 13
Winter Wreath of Summer Flowers, . . . 3 00

TALES FOR THE PEOPLE AND THEIR CHILDREN.

Alice Franklin. By Mary Howitt, ... 38
Crofton Boys (The). By Harriet Martineau, . 38
Dangers of Dining Out. By Mrs. Ellis, . . 38
Domestic Tales. By Hannah More. 2 vols. . 75
Early Friendship. By Mrs. Copley, . . . 38
Farmer's Daughter (The). By Mrs. Cameron, 38
First Impressions. By Mrs. Ellis, . . . 38
Hope On, Hope Ever! By Mary Howitt, . . 38
Little Coin, Much Care. By do. . . . 38
Looking-Glass for the Mind. Many plates, . 38
Love and Money. By Mary Howitt, . . . 38
Minister's Family. By Mrs. Ellis, . . . 38
My Own Story. By Mary Howitt, . . . 38
My Uncle, the Clockmaker. By do. . . . 38
No Sense Like Common Sense. By do. . . 38
Peasant and the Prince. By H. Martineau, . 28
Poplar Grove. By Mrs. Copley, . . . 38
Somerville Hall. By Mrs. Ellis, . . . 38
Sowing and Reaping. By Mary Howitt, . . 38
Story of a Genius. 38
Strive and Thrive. By do. 38
The Two Apprentices. By do. 38
Tired of Housekeeping. By T. S. Arthur, . . 38
Twin Sisters (The). By Mrs. Sandham, . . 38
Which is the Wiser? By Mary Howitt, . . 38
Who Shall be Greatest? By do. . . . 38
Work and Wages. By do. 38

SECOND SERIES.

Chances and Changes. By Charles Burdett, . 38
Goldmaker's Village. By H. Zschokke, . . 38
Never Too Late. By Charles Burdett, . . 38
Ocean Work, Ancient and Modern. By J. H. Wright, ... 38

Picture Pleasure Book, 1st Series, . . . 1 25
" " " 2d Series, . . . 1 25
Robinson Crusoe. 300 Plates, ... 1 50
Susan Pindar's Story Book, ... 75
Sunshine of Greystone, ... 75
Travels of Bob the Squirrel, ... 37
Wonderful Story Book, ... 50
Willy's First Present, ... 75
Week's Delight; or, Games and Stories for the Parlor, ... 75
William Tell, the Hero of Switzerland, . . 50
Young Student. By Madame Guizot, . . 75

www.ingramcontent.com/pod-product-compliance
Lightning Source LLC
LaVergne TN
LVHW020114110826
845151LV00001B/152

* 9 7 8 1 4 2 5 5 4 2 7 5 7 *